MICRO-HYDRO POWER
A guide for development workers

Peter Fraenkel, Oliver Paish, Varis Bokalders, Adam Harvey, Andrew Brown and Rod Edwards

In association with the Stockholm Environment Institute

Practical Action Publishing Ltd
The Schumacher Centre
Bourton on Dunsmore, Rugby,
Warwickshire CV23 9QZ, UK
www.practicalactionpublishing.org

First published 1991\Digitised 2013

ISBN 10: 1 85339 029 1

ISBN 13: 9781853390296

ISBN Library Ebook: 9781780442815

Book DOI: http://dx.doi.org/10.3362/9781780442815

A catalogue record for this book is available from the British Library.

Since 1974, Practical Action Publishing (formerly Intermediate Technology Publications and ITDG Publishing) has published and disseminated books and information in support of international development work throughout the world. Practical Action Publishing is a trading name of Practical Action Publishing Ltd (Company Reg. No. 1159018), the wholly owned publishing company of Practical Action. Practical Action Publishing trades only in support of its parent charity objectives and any profits are covenanted back to Practical Action (Charity Reg. No. 247257, Group VAT Registration No. 880 9924 76).

Contents

ACKNOWLEDGEMENTS

The authors wish to thank the Swedish International Development Authority, SIDA, for supporting the production of this guide and providing the finance for the research work. We also gratefully acknowledge the information and advice provided by the international micro-hydro industry.

Thanks also to Biwater Hydropower Ltd for the use of their drawings in Figures 2.6, 2.7, 7.9 and 7.12, and to Karen Banks for rescuing some of the artwork in Chapter 4.

Foreword

This book is the result of a co-operative project involving I T Power Ltd, ITDG, the Stockholm Environment Institute (SEI), and the Swedish Missionary Council (SMC), and is sponsored by the Swedish International Development Authority. Its origin stems from the needs of SMC field staff who have found much of the information currently available on micro-hydro power to be fragmented and often incompatible.

The Stockholm Environment Institute, which has close working relations with SMC in the field of renewable energy, runs an information programme on renewable energy for development which has resulted in a series of publications and seminars. I T Power and ITDG have substantial experience in the field of micro-hydro power, and much of the material in the guide is based on ITDG's experience over 10 years in Nepal and Sri Lanka where they run training courses in micro-hydro power. I T Power works more broadly in all areas of renewable energy exploitation.

Micro-hydro power is a mature technology which has found applications worldwide in a large number of sites. There are already several publications dealing with the subject, mostly focusing on the technical aspects. Following a series of consulatations with workers in the field, however, we could see that there was a demand for an updated and practical publication specifically addressing the practical needs of development workers. In addition, considerable technological development has taken place during the last few years, particularly as regards electronics and control systems, and this has contributed to make this technology even more realistic and reliable. Our approach has been to complement the technological perspective with other aspects of interest to those contemplating the use of micro-hydro equipment in various locations in the developing world.

This book aims to address the need for updated information, and we hope that the combined experience of four organizations will be of assistance to other workers in the field.

Lars Kristoferson
Stockholm Environment Institute
August 1991

Preface

Micro-hydro power is a collaborative publication involving I T Power Ltd, ITDG (the Intermediate Technology Development Group Ltd), and SEI (the Stockholm Environment Institute). The book was edited and produced by I T Power and is the second in a series, the first of which was entitled *Solar Photovoltaic Products: a guide for development workers.*

The book is intended to assist anyone with some general technical experience, but perhaps limited specific knowledge of micro-hydro systems, to arrive at practical and soundly based decisions on such questions as:

- can a particular stream be utilized to generate power?
- if so, what would be the most appropriate design concept to adopt and what equipment will be needed?
- how can the cost and the economics of such a scheme be determined?
- if it appears to be a viable project, how best should it be designed, procured and implemented?
- after completion, what arrangements are likely to be needed to maintain it in reliable operation?

For the purposes of this book, micro-hydro systems are defined as systems of less than 300kW rated power, which will in most cases also be isolated (or 'stand-alone') systems that are independent of the main electricity grid.

Contact persons at the organizations involved in producing this book are:

Peter Fraenkel
I T Power Ltd
The Warren
Bramshill Road
Eversley, Hants
RG27 0PR, UK

Andy Brown
ITDG Ltd
Myson House
Railway Terrace
Rugby
CV21 3HT, UK

Varis Bokalders
The Stockholm Environment Institute
Box 2142
S-103 14, Stockholm
Sweden

Karl-Erik Lundgren
Swedish Mission Council
Office for International Development
Co-operation
Tegnergatan 34 n.b.
113 59 Stockholm
Sweden

Introduction 1

Water ... the element that knows no rest

Leonardo da Vinci

OF the Sun's radiation that enters the Earth's atmosphere, about one-half is converted into heat directly at the Earth's surface, about one-quarter is reflected back into space, and the remaining quarter is spent in evaporating water, mostly from the oceans. It is this solar energy, at first converted primarily into the latent heat of evaporation of water, that powers the hydrological cycle on which hydro power depends. The water vaporized by incoming solar energy at the Earth's surface rises into the atmosphere until it eventually condenses and precipitates as rain or snow. If it falls on high ground it must flow back towards the oceans from which it came.

The water flowing back towards the sea gradually dissipates its energy as it descends. Most of the descent is sufficiently gradual to make it difficult to extract energy for practical purposes, but in various places where the land-form is suitable, water may be diverted from a river and allowed to fall through a pipe (known as a *penstock)* to a turbine so that the energy can be extracted in a useful form, most commonly as electricity.

Perhaps 9% of the world's technically and economically feasible hydro potential has already been developed and this is sufficient to provide about 23% of the world's electricity from a world installed capacity totalling approximately 375,000MW. A large proportion of this generating capacity consists of large multi-megawatt, and even gigawatt sized schemes, but small-scale 'micro-hydro' is becoming increasingly important.

For example, in the People's Republic of China, which has developed its hydro resources more than most countries, there are over 75,000 small-scale hydro installations (less than 2MW) which generate about 30% of all the electricity used in the rural areas of China. There is clearly great potential for future development of the micro-hydro resource in many other parts of the world.

Hydro power is attractive because it is a renewable energy resource that can never be exhausted and which avoids the pollution associated with burning fossil fuels. However, most of the larger hydro schemes involve massive dams, impounding enormous volumes of water in man-made lakes, in order to provide year-round power by smoothing out fluctuations in river flow. In many cases such schemes are far from inexhaustible because the lakes gradually silt up and will cease to function effectively as inter-seasonal storage reservoirs within a few decades. There are also numerous environmental problems that can result from interference with river flows, and many of the larger schemes have had adverse effects on the local environment.

This book is concerned solely with micro-hydro, or small-scale hydro, which is one of the most environmentally benign energy conversion options available, because unlike large-scale hydro power, it does not attempt to interfere significantly with river flows.

The definition of 'micro-hydro' varies in different countries and can in some cases include systems running to a few megawatts of rated capacity. This book focuses on hydro systems rated from a few hundred watts to a maximum of approximately 300kW capacity. The reasoning is that 300kW is about the maximum size for most stand-alone hydro systems not interlinked to the grid, and suitable for 'run-of-the-river' installation (i.e. without significant damming or the creation of man-made lakes).

Overview of Micro-Hydro 2

2.1 Historical Summary

WATERWHEELS and vertical-shaft Norse wheels or bucket turbines have been in use in many parts of Europe and Asia for some 2,000 years, mostly for milling grain. Two distinct types of watermill were developed: the small, vertical-shaft Norse Mill (Figure 2.1) evolved out of Scandinavia, while the horizontal shaft waterwheels (Figure 2.2) originated in the Mediterranean civilisations.

By the time of the Industrial Revolution towards the end of the Eighteenth century, waterwheel technology had been developed to a fine art, and efficiencies approaching 70% were being achieved for the conversion of water power to mechanical power. Many tens of thousands of waterwheels were in regular use by then, ranging from the original, simple Norse wheels (which are still used today in Afghanistan and the Himalayan region) to sophisticated waterwheels with gear trains, mechanical over-speed protection using fly-ball governors, and scientifically shaped curved iron paddles. The four most common waterwheel configurations are depicted in Figure 2.2.

In England John Smeaton performed serious scientific experiments with waterwheels in the late Eighteenth century. These indicated that overshot wheels (utilizing gravity) were significantly more efficient than undershot wheels (which rely on interaction between the flowing water and the wheel). The British tended to favour overshot wheels after that, but in France serious efforts were made to understand and improve the interaction between flowing water and the blades of a waterwheel. This lead first to the development of the Breast wheel by Poncelet, and then to a much more compact device designed in the 1820s by Benoît Fourneyron, called a hydraulic motor; it was in fact the first hydro-turbine.

Improved engineering and metallurgical skills during the Eighteenth century, combined with the need to develop smaller and higher speed devices to be able to generate electricity without the need for large gear trains, led to the development of turbines. But it was not until the 1880's that hydro turbines were first used to generate electricity for practical purposes.

The compact size and ease of installation of a turbine, and the possibility of using higher

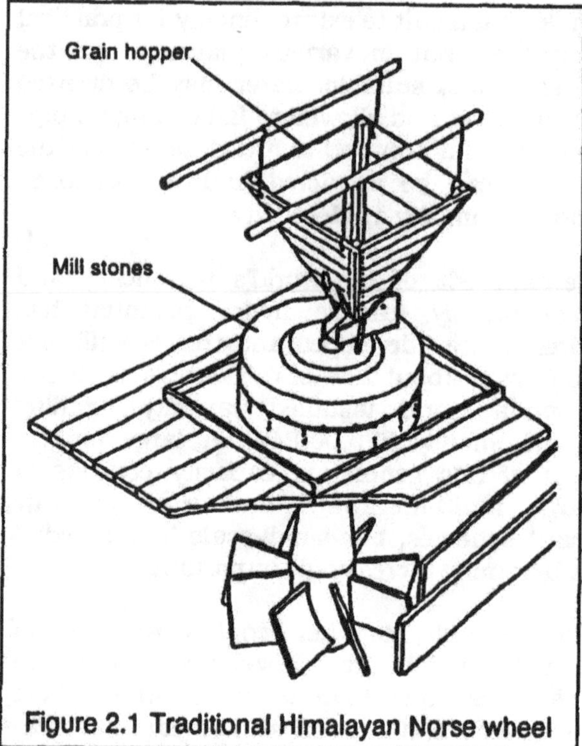

Figure 2.1 Traditional Himalayan Norse wheel

Figure 2.2 Four types of waterwheel

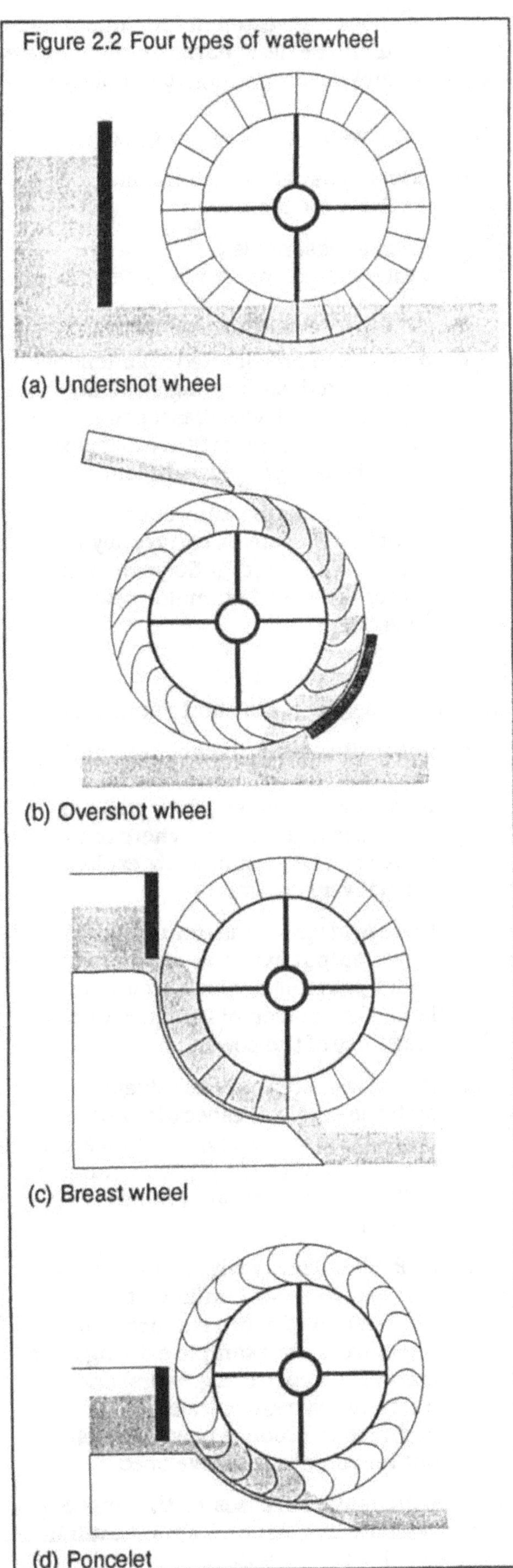

heads and flows to produce greater power and so satisfy the increasing demand for electricity, caused the turbine to take over from the waterwheel almost completely by the end of the century.

Until the 1930s small turbines were increasingly used in Europe and North America. With the development of centralized electricity grids and the economies of scale achieved by large hydro plants, (and with the increasing penetration of subsidized electricity into remote rural areas) there was a steady trend away from small hydro from the 1930s until the 1970s. In fact micro-hydro installations that functioned earlier in the century were allowed to run down as grid connections often linked to fossil-fuel power stations were brought into ever remoter regions. Manufacturers who had produced small hydro-turbines for decades either went out of business or switched their main production capacity to pump manufacturing or other such allied products for which there was a healthier market.

Early micro-hydro was often DC (direct current) because a cruder level of speed control on the turbine-generator set was acceptable compared with more modern AC (alternating current) generation. Thus another reason for the decline of micro-hydro was the increasing standardization towards AC 50Hz (sometimes 60Hz) mains electricity. Most readily-available electrical appliances began to need good quality AC output with only small variations in voltage and frequency. Fast and effective control over the speed and output of a small turbine demanded the use of a sophisticated, precision-engineered control system (usually mechanical speed-sensing with hydraulic servo power to open and close the valve controlling the flow of water to the turbine).

The cost (and power) overheads associated with good quality control systems proved very high for small systems. A control system for a 10kW installation costs much the same as for a 100kW plant and often uses as much as 1kW to sustain itself; this is trivial for a 100kW system but is 10% of a 10 kW output. As a result, the control system for micro-hydro of less than around

15-20kW might cost more than the whole of the rest of the system.

The problem of very costly control systems existed until the recent development of modern solid-state electronic power control systems which provided better and much less expensive technical solutions to this problem, as we shall see.

Since the 'oil shocks' of the 1973-80 period, mains electricity has become increasingly expensive in many countries because the subsidies have been reduced or removed by governments who can no longer afford them. There has been a tendency to look again at power sources that function independently of a fuel supply. This has encouraged new technical developments in the field of micro-hydro power. One of the most important has been the evolution, mainly during the 1970s and 1980s, of electronic control systems that offer for the first time accurate control of even the smallest micro-hydro installations at a reasonable cost.

A lot of work has been done to standardize and 'modularize' small low-head turbines, so that low-head installations can be packaged by manufacturers in a way that reduces the complexity of designing and constructing the civil works (low-head sites tend to be demanding in terms of design skills because no two sites can ever quite use the same arrangements for impounding the water and controlling the flow to the turbine). Numerous novel turbine types have appeared commercially in recent years, such as reversed pumps (an attempt at a cheap turbine by reversing the direction of a standard pump) and various 'bulb' turbines in which the generator is packaged into the turbine hub and submerged in the flow so that the need for a powerhouse can be eliminated.

Many governments are now encouraging the development of micro-hydro power resources in their countries, and there is something of a revival of the industry, with many new turbine manufacturers appearing and some of the older ones enjoying a new expansion. Environmental worries over the greenhouse effect and acid rain should help this to continue, especially since many parts of the world have a significant hydro potential for micro-scale installations of the kind covered by this book. Investment in micro-hydro cannot fail to be less harmful than money spent on fuel-burning energy technologies.

The main advantages of hydro power are:

- power is usually continuously available on demand
- given a reasonable head, it is a concentrated energy resource
- the energy available is predictable
- no fuel and only limited maintenance are required, so running costs are low (e.g. compared with diesel power) and in many cases imports are displaced to the benefit of the local economy
- it is a long-lasting and robust technology; systems can readily be engineered to last for 50 years or more without the need for major new investment

Against these, the main shortcomings are:

- it is a site-specific technology, and sites that are well suited to the harnessing of water power and are also close to a location where the power can be economically exploited are not very common
- there is always a maximum useful power output available from a given hydro power site, which limits the level of expansion of activities which make use of the power
- river flows often vary considerably with the seasons, especially where there are monsoon-type climates, and this can limit the firm power output to quite a small fraction of the possible peak output
- lack of familiarity with the technology and how to apply it inhibits the exploitation of hydro resources in some areas; for example existing river control structures (e.g. weirs) and water supply reservoirs could be exploited to produce power as well, but are often left undeveloped.

One of the main purposes of this guide is to help overcome this last and important inhibiting factor: lack of knowledge.

2.2 General Principles

The basic physical principle of hydro power is that if water can be piped from a certain level to a lower level, then the resulting water pressure can be used to do work. If the water pressure is allowed to move a mechanical component then that movement involves the conversion of water energy into mechanical energy. Hydro-turbines convert water pressure into mechanical shaft power, which can be used to drive an electricity generator, a grain mill, or some other useful device.

Head

The term *head* refers to the actual height that the water falls through. Figures 2.3 and 2.4 illustrate this distance in a medium-head and a low-head hydro installation.

Power

Sometimes people confuse the words 'power' and 'energy'. Power is the *energy converted per second* i.e. the rate of work being done. Energy is the total work done in a certain time.

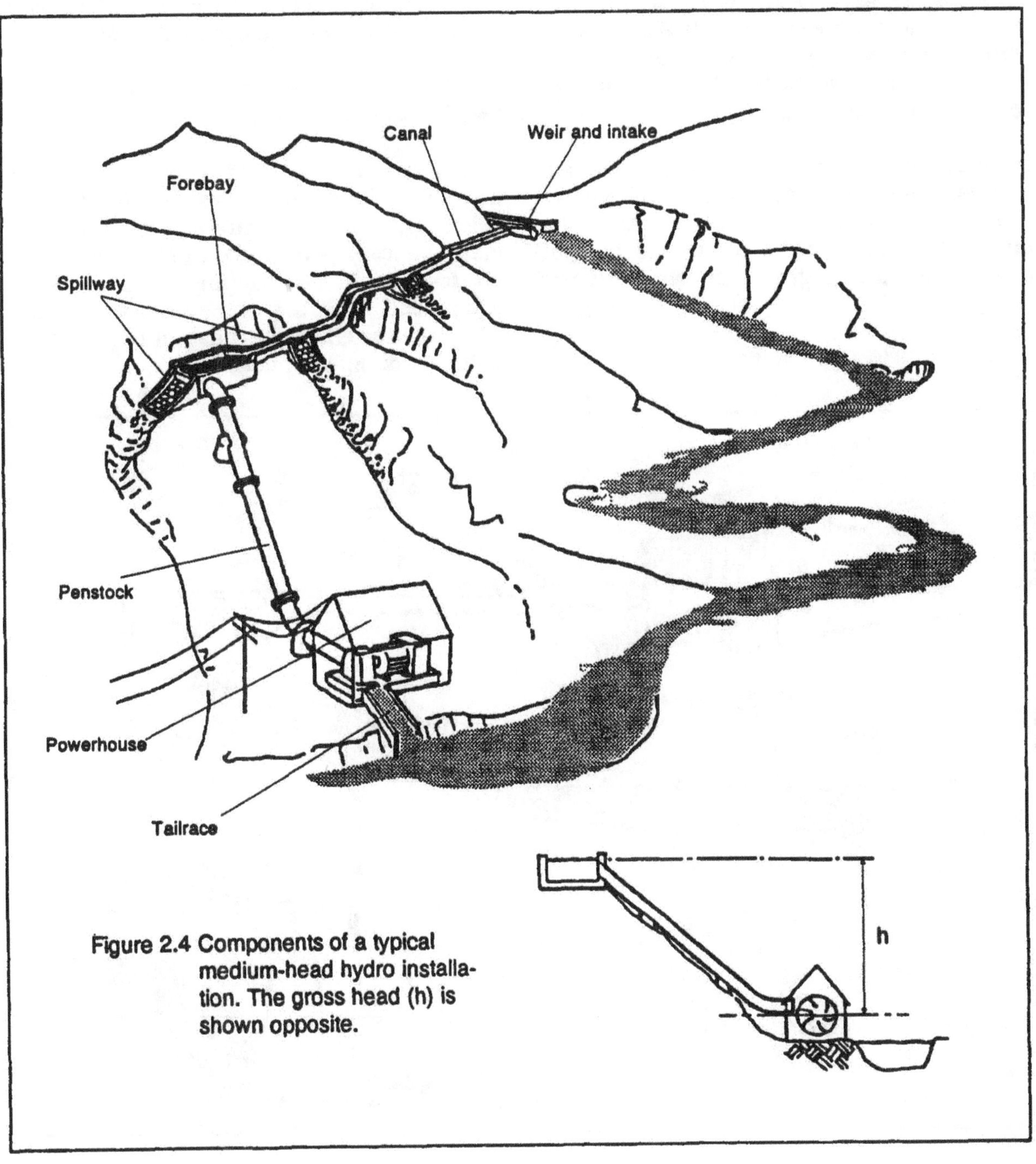

Figure 2.4 Components of a typical medium-head hydro installation. The gross head (h) is shown opposite.

The theoretical power P available from a given head of water is in exact proportion to the head H and the flow rate Q.

$$P = Q H \times \text{a constant}$$

The constant is in fact the product of the density of water and the acceleration due to gravity. In SI (Systeme Internationale) units, if P is measured in watts, Q is in cubic metres per second, and H is in metres, then the actual gross power of a flow of water is:

$$P = 1000 \times 9.81\ Q H \quad \text{Watts}$$

where the density of water is $1000kg/m^3$ and the acceleration due to gravity is $9.81m/s^2$. It follows that for a given power output, the greater the head available the less the flow required, and vice versa.

In practice, as with all real-life devices, hydro-turbines are not perfectly efficient in converting the energy of falling water into mechanical shaft power, so the real output can be calculated from the simple basic formula:

$$P = k\ 9810\ Q H \quad \text{Watts}$$

where k is a number less than one which expresses the efficiency of the turbine (e.g. if the turbine converts 70% of the energy of the water flow into shaft work, then it is 70% efficient and $k = 0.7$). For example, given a head of 20m, a flow of 40l/s (or 0.04 m^3/s) and a turbine efficiency of 70% we can estimate the output as:

$$P = 0.7 \times 9810 \times 20 \times 0.04$$

$$P = 5494 \text{ Watts (or 5.5kW)}$$

Since watts and litres are rather small units to work with except with the smallest of micro-hydro plants, it is easier to work with kilowatts (kW) and cubic metres. In summary:

There are 9.8kW available from each cubic metre per second of water falling through 1 metre.

Therefore, given knowledge of the head, the flow rate, and some idea of the turbine efficiency, it is immediately possible to estimate the likely power output. This fundamental formula applies to any kind of hydro equipment, be it a turbine or a waterwheel (although the efficiency constant may be quite different for different devices).

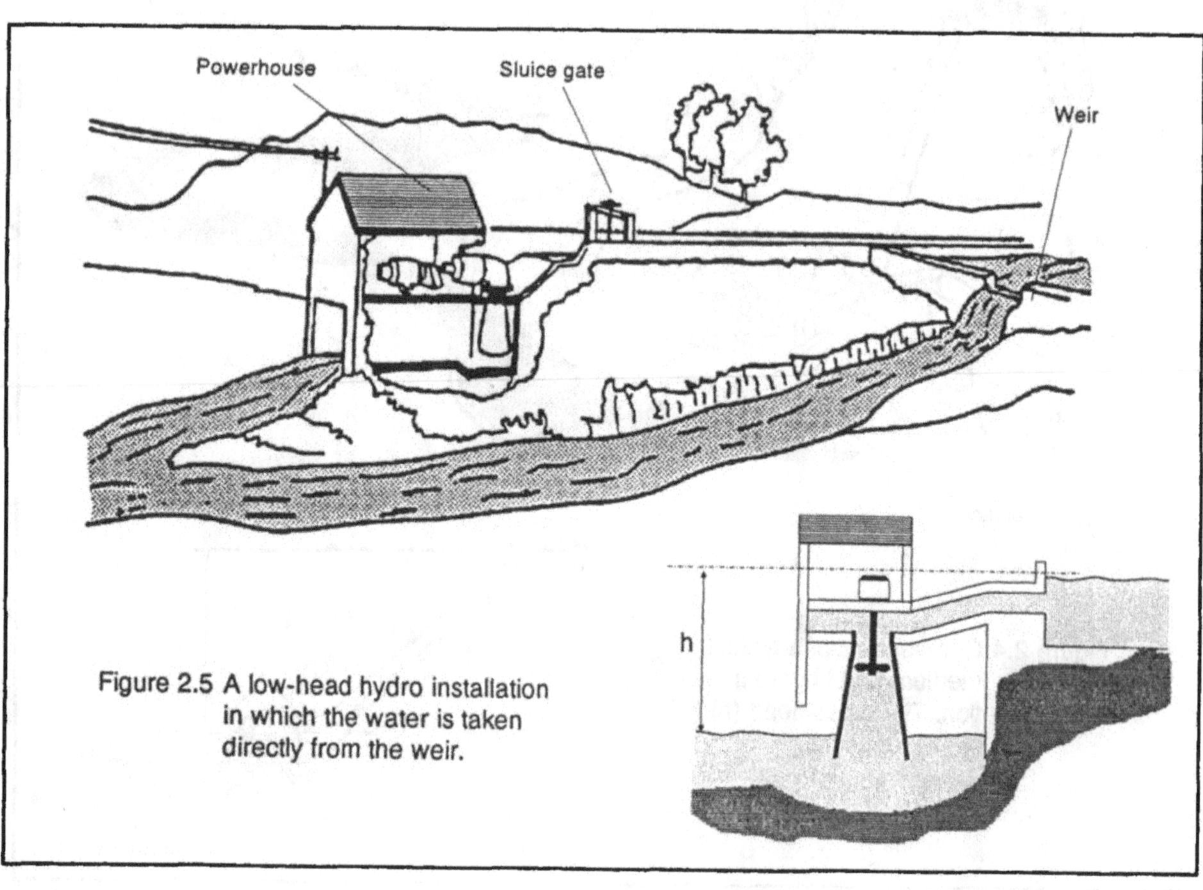

Figure 2.5 A low-head hydro installation in which the water is taken directly from the weir.

2.3 Hydro Power Equipment

The kind of turbine that is appropriate for different purposes depends on the flow and the head. Certain types of turbine are best suited to low heads, others for high heads. For instance waterwheels are only suitable for low heads.

Figure 2.5 is a family of curves showing the power produced for a wide range of heads and flows when the overall efficiency is 70%. This is close to the typical efficiency of many small hydro-electric systems but the result can easily be corrected for other efficiencies. For example 60% efficiency would produce a power output 6/7ths of that indicated. The curves can provide an estimate of the likely output from any hydro plant.

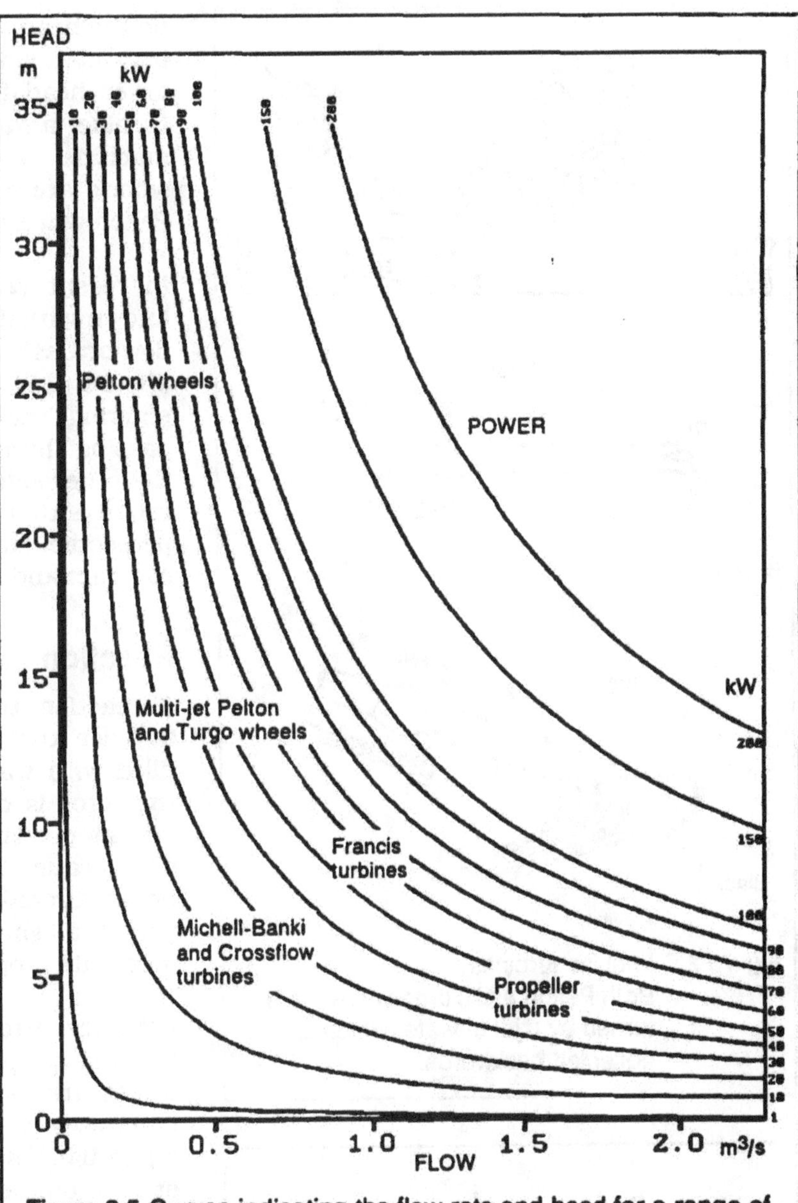

Figure 2.5 Curves indicating the flow rate and head for a range of power outputs, assuming an overall efficiency of 70%.

The main reason different types of turbine are used at different heads is that electricity generation requires a shaft speed as close as possible to 1,500rpm (revolutions/minute) to minimize the speed change between the turbine and the electricity generator (or *alternator*). The speed of any given type of turbine tends to decline in proportion to the head, so low-head sites need turbines that are inherently faster under a given operating condition. Similarly a high-speed type of turbine could cause problems at higher heads if for some reason the load came off and it 'ran away' to its maximum overspeed condition, perhaps destroying the alternator. Also as more flow is needed at low heads to achieve a given power output, low-head turbines need to pass a much greater flow of water than a high-head machine.

There are two main families of turbines: impulse turbines and reaction turbines.

Impulse turbines

In an impulse turbine, the pressurized water is converted into a high-speed jet by passing it through a nozzle. The jet strikes the specially shaped buckets or blades of the turbine rotor, causing it to rotate.

Typical impulse turbines are the Pelton wheel and the Turgo wheel (Figure 2.6). Because an impulse turbine uses a jet of water in air, its casing is at atmospheric pressure and serves more as a splash guard than as a container for the water.

Impulse turbines are generally used in medium to high-head applications. The limits

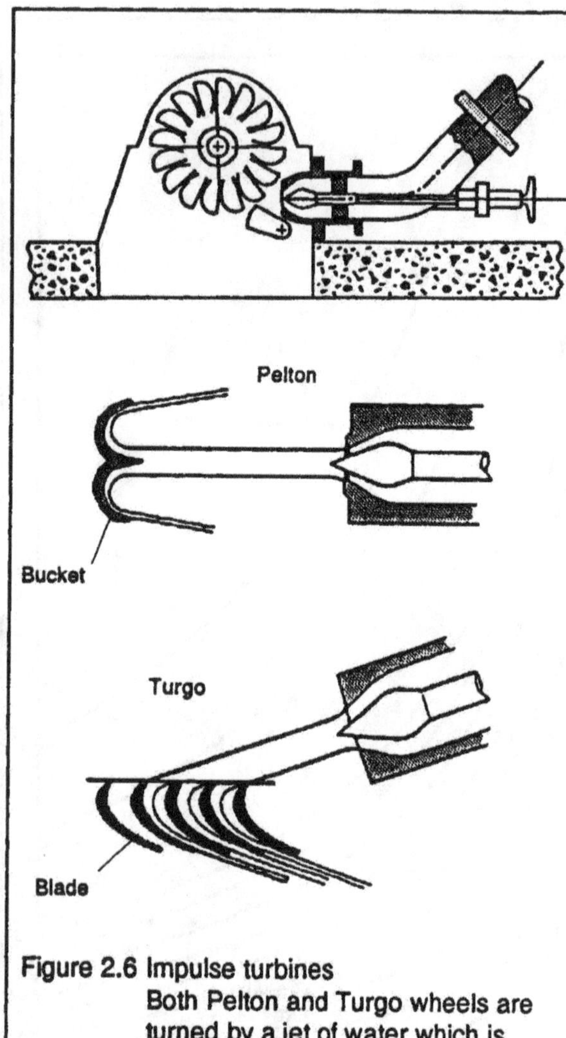

Figure 2.6 Impulse turbines
Both Pelton and Turgo wheels are turned by a jet of water which is deflected backwards.

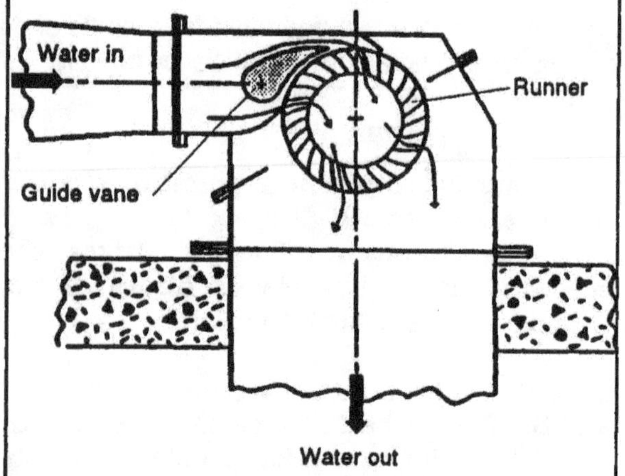

Figure 2.7 Crossflow turbine
The water passes through the centre of the runner, striking the blades twice.

on the head are also a function of turbine size because a small Pelton wheel can operate at a lower head than a large one since a small turbine always runs faster than a larger machine of the same type. In fact Turgo wheels are more suited to lower heads than Pelton wheels.

In recent years a type of impulse turbine known variously as the Michell-Banki, Cross-flow or Ossberger turbine has become popular (Figure 2.7). This turbine has the water entering the runner as a rectangular jet, passing through the runner radially and out the other side. An advantage of this configuration is that the water enters and discharges perpendicularly to the shaft, which allows a compact and simple installation.

Reaction turbines

A reaction turbine is quite different to an impulse turbine in that it runs completely filled with water. The flow of water through the rotor is deflected in such a way that it creates pressure differences across the blades which cause them to rotate (similar to the lift on an aircraft wing, or the force produced to propel a ship by the pressure differences across its propeller blades).

Propeller turbines, in which a rotor shaped like a ship's propeller is immersed in a flow of water, are used at low heads. As Figure 2.8 illustrates, they can either have water entering radially, spiralling in from above as in (a), or be mounted in a tube (b) with fixed guide vanes to counteract the rotation the rotor gives to the water. A variation of the tube turbine which is much more compact but complicated is the bulb turbine (c) where the entire generator is immersed in a water-tight pod which carries the rotor. The more advanced propeller turbines (known as Kaplan turbines) have blades which can be rotated about their point of attachment to the hub so that they cut the water at different angles depending on the flow rate through the turbine, thus maximising the efficiency.

Medium heads were traditionally handled by the Francis turbine (Figure 2.9). This is a reaction turbine with a snail-shaped casing. The water enters tangentially and is directed radially inwards to flow into a rotor which is turned by the water so as to discharge it

Figure 2.8 Propeller turbines

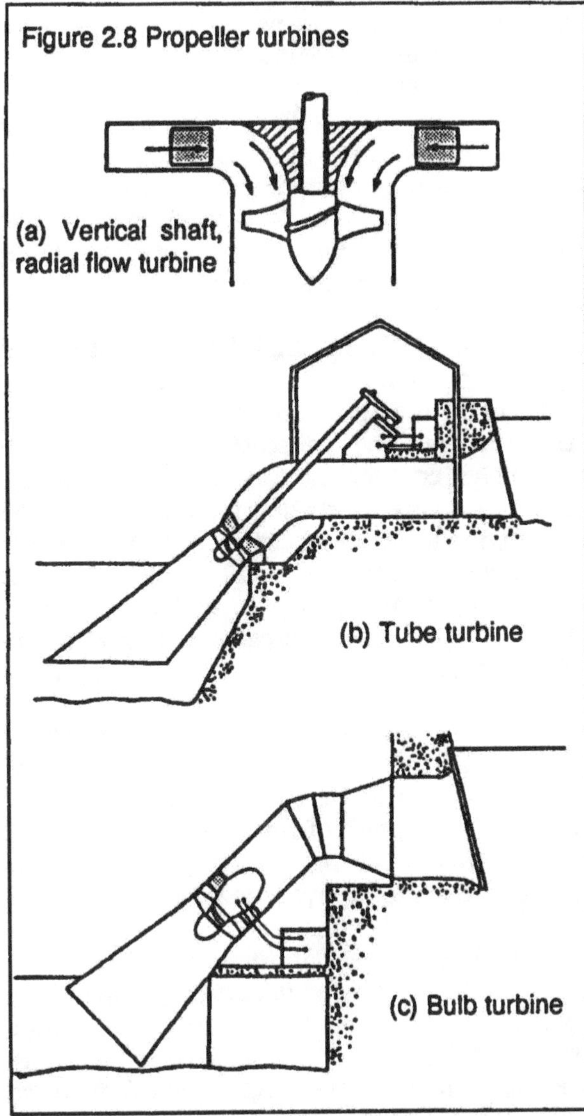

Table 2.1 Efficiencies of hydro systems

TYPE OF DEVICE	EFFICIENCY (shaft power)
Undershot waterwheel	0.25 - 0.40
Overshot waterwheel	0.50 - 0.70
Breast waterwheel	0.50 - 0.60
Poncelet waterwheel	0.40 -0.60
Vertical shaft watermill	0.20 - 0.35
Impulse turbine	0.70 - 0.87
Crossflow turbine	0.60 - 0.80
Reaction turbine	0.65 - 0.90

axially. The Francis turbine is complex to manufacture due to the difficult shape of castings required, and therefore tends to be expensive.

Waterwheels

In addition to turbines, which are normally used to drive electricity generators, there are other devices that can utilize water power, usually for other purposes. The traditional device is the waterwheel (Figure 2.2). There are four principal types: (a) undershot, where the water passes underneath the wheel, for very low heads of less than 1m, (b) overshot, in which the water falls into buckets on the rim of the wheel and the weight of the water carries the wheel around, (c) the breast wheel, essentially a less efficient version of the overshot wheel but which requires the same effort to build, and (d) the Poncelet, a refined version of the undershot wheel in which the water is forced through a narrower opening at a higher velocity, and the buckets are curved so as to capture the water energy more effectively.

The main reason waterwheels are rarely used for electricity generation is that they rotate much more slowly than a turbine and therefore need a large and expensive (and troublesome) speed-increasing mechanism to drive an alternator at the required 1500rpm.

The Norse wheel or Bucket turbine is a form of vertical shaft turbine that has been used in a crude form for hundreds of years (Figure 2.1). In improved form with metal curved buckets it approximates to a form of Turgo wheel for low heads and can be quite efficient.

Efficiency

Table 2.1 provides an indication of how efficient each hydro power device is in converting the energy of falling water into shaft power.

The efficiency factor of 0.7 (or 70%) used in Figure 2.5 is approximately correct for typical small hydro-turbines combined with alternators. Although a good turbine may be 80% efficient or better, losses in the alternator and any speed-changing mechanism will always reduce the overall efficiency.

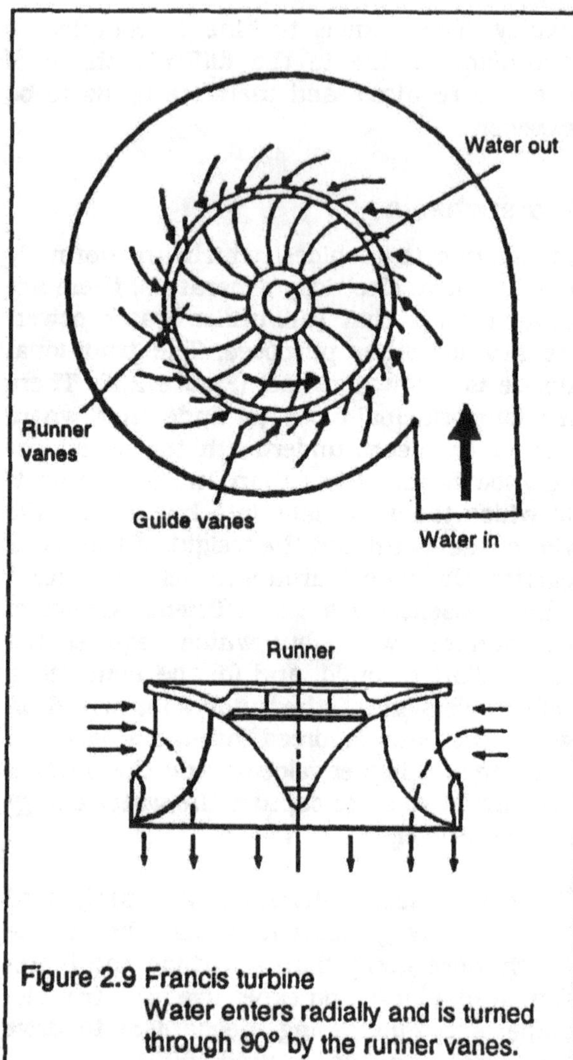

Figure 2.9 Francis turbine
Water enters radially and is turned through 90° by the runner vanes.

2.4 The Components of a Hydro System

Figures 2.3 and 2.4 respectively illustrated a medium-head and a low-head micro-hydro installation. In either case water is taken from the river by diverting it through an intake at a small weir. The weir is a man-made barrier across the river which is built to keep the water level at that point at a constant level to maintain a continuous flow through the intake. The intake should be placed as high as possible above the turbine so as to maximize the head and hence the power.

The intake is usually protected by a rack of metal bars (a *trash rack*) which filters out water-borne debris such as grass or bits of timber. Before descending to the turbine, the water passes through a settling tank or *forebay* in which the water is slowed down sufficiently for suspended particles to settle out on the bottom.

In the medium-head or high-head installations water may be carried to the forebay by a small canal or *leat*. Low-head installations generally involve water entering the turbine almost directly from the weir.

Medium-head and high-head installations have a pressure pipe, known as a *penstock*, which conveys the water from the forebay to the turbine. The penstock pipe can be an expensive component for high-head installations and care is needed to size it correctly; too large a diameter can be unnecessarily expensive, while too small a diameter can cause significant energy to be lost to friction in the pipe, which means a reduction in effective head at the turbine.

All installations need to have a valve or sluice gate at the top of the penstock which can be closed when the turbine needs to be shut-down and emptied of water for maintenance. When this valve is closed, the water is diverted back to the river down a *spillway*.

Effective speed regulation and control is important in electricity generating systems to ensure that the voltage and frequency remain constant. In some cases a valve is automatically driven by a governing mechanism which adjusts the flow to the turbine to meet variations in power demand. If the turbine slows down, the valve is opened to let more water through, and vice versa. In other cases the turbine always runs at full power and control is achieved by adjusting the electrical power output rather than the water power input. In this situation excess electrical power is switched to a ballast load. This has to be capable of absorbing and dissipating the full rated power of the turbine at times when there is zero demand.

Water leaving a reaction turbine discharges through a *draft tube* which is carefully shaped to slow down the water so that it enters the tailrace with minimum energy. Slowing down the water increases the pressure drop across the turbine and hence the available power. Impulse turbines usually need little more than a closed casing to prevent splashing as the water falls out below the turbine.

To produce electricity, the turbine drives an alternator, either directly (if the shaft speed is equal to the desired alternator speed) or through a mechanical transmission which steps up the shaft speed to meet the alternator speed. The alternator should be equipped with sensors to control it and to shut down the system in the event of over-speed or over-load.

The electricity output will either be single-phase Alternating Current (AC), for the smallest installations generally of less than 10kW, or three-phase AC. Usually in the power ranges of interest for this guide it will be generated and transmitted at the same voltage as the user needs (often 240V and 50Hz). Where larger power levels and longer transmission distances are involved, transformers are sometimes used to step up the voltage, which reduces the transmission losses.

Micro-hydro for battery charging

It is possible to use the smallest streams to run very small hydro plants with electrical outputs of under 1kW. In such cases it is not economic to transmit the electricity to a distant load because the cost of the cables will be out of all proportion to the value of the power. However if the power is generated at 12 or 24 Volts DC (Direct Current), it can be used for charging standard lead-acid batteries. A battery can then be transported to where the power is needed and will be sufficient to power a light or two, or perhaps a small radio.

A handful of manufacturers, mainly in North America, produce miniature turbines of this kind for battery charging. They are relatively inexpensive and easy to install, but of course produce only a small output. It is also possible to improvise a turbine of this kind, for example by using an alternator from a lorry (with diodes to rectify its output from AC to DC). It should be noted that motor vehicle alternators will not have a very long life in this role as they are not designed for continuous operation.

If the main energy requirement is small-scale lighting for houses which are some distance from a hydro-site, it may be easier for a larger micro-hydro installation to use a battery charger to provide 'portable electricity' for more distant consumers rather than supply by cable. Each household would normally have two batteries, so that one is in use while the other is recharged.

Preliminary Studies 3

THIS chapter covers the steps to be taken at the beginning of a micro-hydro project, in order to assess whether the idea is worth carrying forward. This will usually involve a pre-feasibility study, a feasibility study, and a design study.

The *pre-feasibility study* typically considers a variety of alternatives and uses approximate data. The *feasibility study* narrows the choice down to one or two options. It looks more closely at costs and time scales, and requires more accurate data to be collected. A feasibility study will usually be the document which is considered by funders, community organisations, the management or other decision-makers. If the project continues past this stage there will then be a *design study*, which is the point at which specifications, orders and tender documents are prepared. For many smaller projects these stages may merge together, but the pre-feasibility study is always needed.

3.1 The Resource

The first question to ask is what order of hydropower potential is available. Consider the highest head options first because they will usually cost less per installed kilowatt.

For pre-feasibility, the most common approaches for determining head are large-scale maps for high-head sites (>60m), pressure gauge and tube for medium-head sites, and hand-held levels for low-head sites (<15m). The most common approaches for flow determination are basic hydrological models or simple flow measurement spot checks. These techniques are discussed in Chapter 4.

The rainfall and catchment area feeding a site can always be identified to within about 50%, which is adequate at this stage. This will give the total potential flow past the proposed intake. After an allowance has been made for evaporation, the average flow can be estimated along with the potential power of the site at this flow. A typical Flow Duration Curve for the region (Section 4.2) will provide information on how the flow varies through the year, but spot flow measurements should be taken at the site to back up this data. Once an adequate resource has been identified, an assessment needs to be made of the likely demand for electricity.

3.2 The Demand

This is an area of micro-hydro design in which there are no certain answers. Good estimates and realistic predictions are the best that can be achieved. Perhaps the best way to introduce this topic is to look at some examples.

Consider first an isolated school with dormitories, kitchens and workshops. The school has a diesel generator which is run in the evenings. The first pointer is the capacity of the generator. The next step is to look at the energy used. There may be an electrical energy meter, records from which can provide answers at once, but it is usually necessary to make a record of a typical day or week. One or more ammeters and voltmeters may have to be introduced into the system to measure the power being consumed. A cross check can be made from fuel consumption (1 to 2 kWh/*l* is typical for small generators). In this way the present electrical consumption and peak power can be found. There is a further aspect to explore, however, because the introduction of a hydro system may encourage new uses for its cheap power, perhaps cooking, refrigeration, etc. replacing their current fuels such as wood or diesel. These possibilities need to be identified and allowed for in the estimate of demand.

The second example is a village which has never had an electricity supply. First, it is necessary to map out the positions of dwellings to find the lengths of power lines required. This may rule out some dwellings on connection-cost grounds. Then some idea of lighting demand can be found from the sizes of the houses and the current usage of candles or kerosene lamps. People are generally willing to pay as much or more for electric lighting than for kerosene, but they expect a far better quality system. Interviews may be carried out to find out what people can pay and what other uses for power exist now or in the future. Any local industries, particularly those using generators or large quantities of wood fuel , need to be examined in detail to assess their likely demand.

In all cases it is usual to plot out daily load curves and, if the supply is a constraint, also to plot out seasonal curves to see whether, for example, there is more demand in wet weather, justifying the choice of a machine which only works at full power for part of the year.

3.3 Preliminary System Sizing

Once supply and demand have been estimated, some attempts at choosing a system size can be made. All kinds of constraints must be considered. There may, for example, be a top limit on the capital available, despite plenty of available power and existing demand. Or there may be legal limits to system power for private schemes. Either supply or demand is usually becoming dominant by this stage of the pre-feasibility. There is no great problem if the supply exceeds the demand. The more difficult case occurs where demand exceeds supply. It then has to be decided whether power which is not 'firm' (not available all year) is sufficiently useful, or if the demand can be broken up into different areas for which hydro may only serve some pre-defined needs.

The easiest 'demand' to deal with is when connecting to a large grid, as a grid can absorb power at any time. Non-firm inputs to the grid are less useful, however, and many utilities or grid operators include penalties for non-supply, reducing the effective price obtained per unit.

Eventually it will be possible to outline a few acceptable schemes within the physical possibilities, capital limits and legal constraints. At this point the economics of the situation must be brought into the picture.

3.4 Basic Economic Analysis

For the purposes of prefeasibility, a few major economic pointers are sufficient. These are:

- capital cost
- annual revenue
- operating and maintenance costs

It should be noted carefully that in the early stages of planning a hydro scheme it is common for undue optimism to lead to an underestimate of the costs and an overestimate of the energy available.

Capital cost

The cheapest micro-hydro systems are locally built and purely for mechanical applications, such as milling. They can cost as little as US$200/kW. The most expensive are usually 'turn-key' electrical systems in remote areas which use high quality imported western equipment and contractors. These may reach as much as US$10,000/kW.

Most practical electricity generating micro-hydro systems fall within the range US$1,000 to US$4,000, and this can be used as a guide in the early stages of a pre-feasibility study. When a few sites are being considered more carefully, budget prices of major components are needed.

The major components of a hyro scheme may be sub-divided as follows:

- engineering (design, supervision, etc.)
- weir and intake
- channel and forebay
- penstock
- turbine , with governor, generator, etc.
- powerhouse
- transmission lines

Other items such as insurance, commissioning costs, site clearance, access tracks, distribution cables, meters, etc. should be considered but can often be shown as a 20% blanket 'other costs and contingencies' at least until the feasibility study stage.

Annual revenue

This can only be an estimate and will be based on putting a value on the demand. For electrical systems the national grid charging system could be adopted, but this is nearly always subsidized for small rural consumers and may be a difficult target to equal. Perhaps the most business-like rate to charge should be based on the actual cost of the micro-hydro system and some assumption on the likely load factor (load factor is the fraction of available energy that is actually consumed and earns revenue).

Other guides for charging or valuing power might be the cost of using diesel, or the value of the fuel (say wood and kerosene) that the hydro is replacing. As an approximate guide, electricity is rarely sold for less than US$0.04/kWh. Large-scale and micro-hydro tend to cost in the region of US$ 0.1-0.2/kWh and diesel generated power often costs around US$0.40/kWh.

For mechanical systems, saw mills, flour mills, etc. the system is usually analysed as a business. The mill can process so much flour, each kilogram processed brings a certain revenue, and so on, based on giving an appropriate return on the investment.

Whichever method of assessing income is used, the estimate should also include the possibility that a series of dry years may occur immediately after the plant is finished.

Operation and maintenance costs

Few European hydro-electric generating systems of less than 200 kW are staffed continuously. In many developing countries a guard/ operator is employed, and in some cases government regulations require that excessive numbers of staff 'mind' 50 kW sets. This latter requirement is never economic and has been used as an argument against micro-hydro in the past. It is therefore worthwhile exploring ways of minimizing staff, especially on smaller sets. In tea factories for example, locating the powerhouse in the factory may lose some metres of head, but may mean that no extra staff are needed.

Maintenance costs are often assumed to be 3% of the original capital investment per year. In practice it is downtime (lost income due to an idle system) that costs more than the maintenance itself, and it is important to allow from two days to a week or so per year in total for this, depending on location, quality of equipment, calibre of skilled staff, etc.

Economic decisions

The economic viability of a micro-hydro scheme depends upon how much it will cost to set up (the *capital cost*) and how much money it will earn when it is running. Much of the revenue from selling the electricity will have to be spent on the operation and maintenance of the scheme. The money left over each year (the *net annual earnings*) will be used to pay back the money borrowed when setting up the scheme. An estimate of capital cost and net annual earnings permits a number of indicators to be produced. For comparative work, the *simple payback period* is often preferred:

$$\text{Simple payback period} = \frac{\text{capital cost}}{\text{net annual earnings}}$$

Most business investments, micro-hydro schemes included, usually need to pay back if possible within three to four years and certainly in less than six to eight years, i.e. the capital cost needs to be three or four times (maximum six to eight) the net annual revenue expected from the system.

3.5 Environmental Considerations

Although micro-hydro power is one of the most environmentally benign methods of generating electricity, it is not totally without environmental problems. These need to be considered during the planning stage and solutions found which are not only practical but will also meet government or river authority regulations.

Some of the main environmental problems are:

Fish
Hydro-installations on rivers populated by migrating species of fish, such as salmon or trout, are generally subject to special requirements. Fish must not be ingested into the turbine (so the mesh of the trashrack must be fine enough), and there must be a water passage by-passing the hydro-plant at all times so that fish can migrate up or downstream. To allow fish to pass upstream sometimes requires the construction of a *fish ladder*, which is usually a series of pools one above the other, with water overflowing from the higher ones to the lower ones, so that fish can jump up from one pool to the next.

Irrigation
If the river is used for irrigation it is vital when designing and sizing the hydro-system to allow for the water removed from the river for irrigating crops. Finding the necessary information may require discussions with the local farmers

Erosion
The river banks or any canals that are to be dug may be vulnerable to erosion as a result of the hydro-project. Care is needed to ensure that erosion is minimized, and that if it does occur, there will be no risk of flooding to nearby farmland or property. Consideration should also be given to the effects of either the penstock rupturing or the entire weir being washed away; would habitations or even lives be put at risk? Fortunately most micro-hydro systems involve little or no impounded water so the danger of catastrophic flooding will not normally arise.

Disease
In some tropical areas water-borne diseases can be prevalent and may pose a risk to the construction team and operators of the system. Fortunately the most serious water-borne diseases are found in places where the water is either still or slow-moving. In most situations where hydro power extraction is contemplated there is unlikely to be much danger unless the water is flowing out of a nearby lake or other area containing stagnant water.

Electricity
Every electrical supply poses obvious hazards, and the safety of people and animals must be treated very seriously. Generally, if the electrical installation is built to accepted national and international standards such dangers should be eliminated.

Penstocks and Cables
Surface-mounted penstocks and transmission cables may conflict with paths and tracks, or with farmers' fields, so decisions on the layout of the penstock and the power lines need to take account of possible conflicts. In some cases burying both underground is the required solution, but this is obviously subject to ground conditions.

Hydrology and Site Survey 4

4.1 Introduction

THE first requirement for a successful hydro scheme is to find the best possible site. This will be a site which is conveniently near the demand and where the product of flow and head are likely to be maximized. When a choice of placement is possible it is essential to obtain maps and regional records of rainfall and water flows in streams and rivers (known as *hydrological records*). Flow and head need to be determined and of these, flow, which is a variable, is the most difficult to determine. Records may be available from the government if appropriate studies have been completed. Section 4.2 presents a method for estimating the water intake of a micro-hydro scheme as it varies through the year. It involves a simple approach to hydrology, which has proved valid in certain areas but is not necessarily valid everywhere.

The reason that the hydrological study is both the most important and the most difficult part of the hydro-design process is that surface water flow varies through the year in a complex manner. It is a function of the hydrological cycle, which is depicted in a simplified form in Figure 4.1.

In a large-scale hydrological study, a frequent cause of surface flow variation through the year is the large release of water in spring due to snow melt. It is often assumed that the flow variations occurring in micro-hydro catchment areas will be similar to the variations in larger catchment areas, and as a result hydrological data relevant to full-scale hydro is extrapolated for micro-hydro design purposes. This can be a false assumption which will lead to design errors. One reason

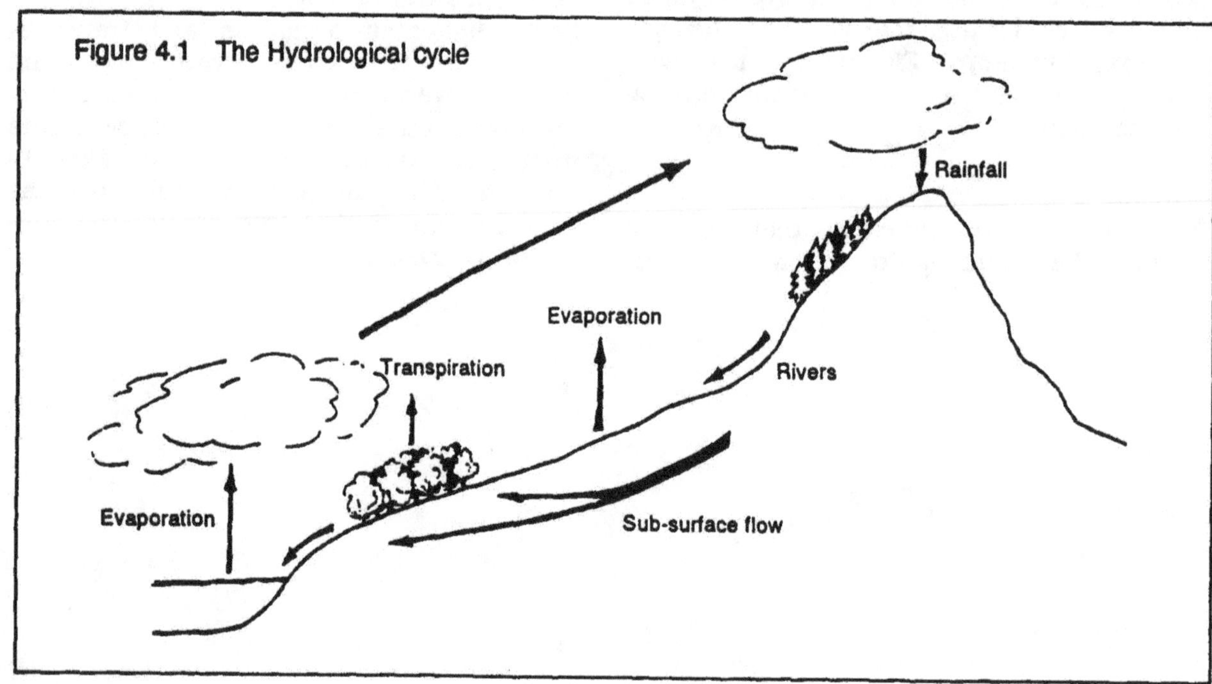

Figure 4.1 The Hydrological cycle

to be wary of large catchment area hydrology data is that the ground cover in small catchment areas may be quite different to that in large catchment areas, so the evaporation characteristics will not be comparable. Another reason is that micro-hydro catchment areas are not usually fed from high altitudes where snow melt takes place, and therefore do not experience the high summer flows often associated with snow melt in large catchment areas.

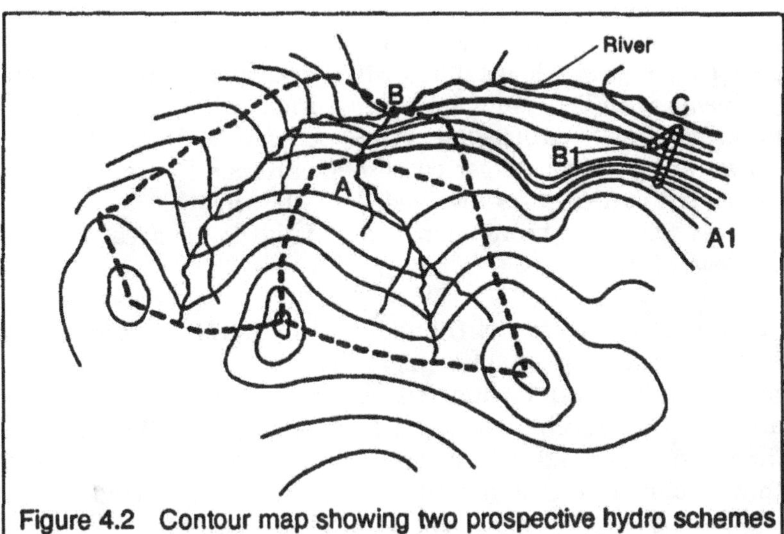

Figure 4.2 Contour map showing two prospective hydro schemes

It should be added that there *are* areas where ground cover is relatively consistent between large and micro-hydro catchment areas, and where snow melt never occurs. In such areas extrapolation from large-scale hydrology to micro-hydrology is a very quick and useful method of proceeding and the prospective micro-hydro user might start by seeking data from any large-scale hydro studies produced by the government, and scaling them down accordingly.

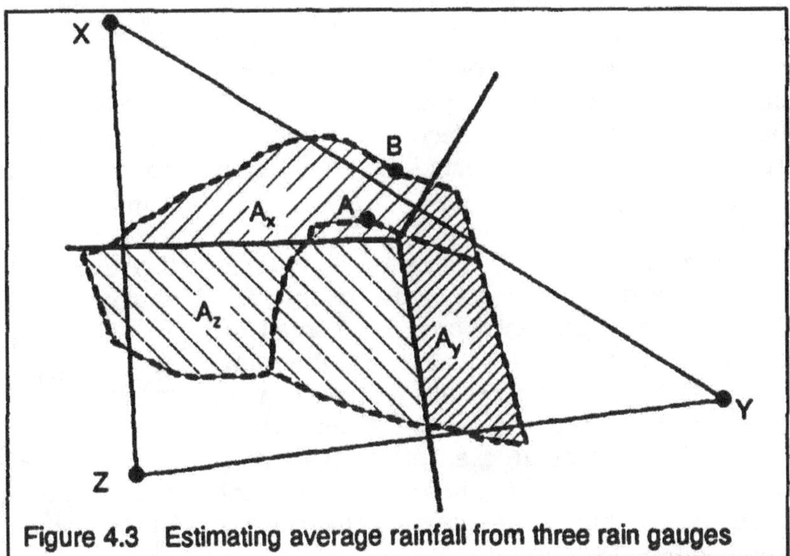

Figure 4.3 Estimating average rainfall from three rain gauges

4.2 Comparison of Catchment Areas

Careful examination of a contoured large-scale map may show that a number of possibilities exist for the design of a hydro scheme. Consider Figure 4.2 which shows a section of a map: two possible micro-hydro schemes have been outlined, both making use of a turbine placed at point C. Scheme A draws water at point A, then channels it along a contour line to point A1. The relatively small drop needed for the channel is ignored in this case because the overall head is large. With lower overall heads the slope needed for a canal can be significant and needs to be allowed for. A1 represents a proposed forebay tank feeding a penstock which drops to the turbine at C. B is an alternative proposal for an intake site. The channel feeding water to the penstock at B1 is shorter than channel A, and covers ground which is less steep; the penstock is about half as long.

Would it be better to draw water from point A or point B? Since obtaining the maximum head is very important in a hydro scheme, an intake sited at point A could be preferable to one sited at B. On the other hand, the penstock for scheme A is longer and so more expensive. The length of channel A is also greater, and because of the steep slope may be expensive to build and maintain, so raising the total cost of the scheme. If necessary a geological survey will establish whether or not problems with land slips or storm runoff will make channel A unfeasible. Also, scheme B draws water from a larger catchment area as extra tributaries are involved, so the increased flow may more than compensate for the lower head.

To analyse the merits of the different options it is necessary to consider the area in which rainfall collects and runs towards points A and B. The perimeters of these areas are drawn in Figure 4.2 with dashed lines. Notice that catchment B includes all of catchment A.

Finding the average daily flow (ADF)

The average daily flow (ADF) of the river is calculated from the yearly flow. An estimate of average yearly flow at points A and B can be made from existing records of rainfall, such as information collected at nearby rain gauges, as explained below.

Suppose that three rain gauges are situated somewhere near the catchments of interest. Any such rain gauges should be located on the map; if not, mark them on the map yourself. Any number of rain gauges can be used in the manner described below, three being taken only as an example. This method can even be used for gauges quite distant from the catchment, perhaps at the nearest meteorological station.

Figure 4.3 shows the same area as Figure 4.2 but with the contours omitted, for clarity. Three rain gauges are located at points X,Y, and Z. Because Z is nearer to the catchments, and Y and X are further away, Z can be considered to have more influence on the average rainfall in the catchment. One way of weighting the average is to draw the polygon shown in Figure 4.3 (in this case a triangle) linking X, Y and Z, and then bisect each side, so referring an area of the catchment to each rain gauge. If rainfall at each gauge is l_x, l_y and l_z mm/year, then average rainfall in the catchment (in mm/year) can be estimated as:

$$l_{av} = \frac{A_x}{A_{tot}} l_x + \frac{A_y}{A_{tot}} l_y + \frac{A_z}{A_{tot}} l_z$$

where $A_{tot} = A_x + A_y + A_z$

The simplest method for estimating areas from the map is to use squared tracing paper and count the squares in each area. If the scale value of the map is used to estimate the area of each square then the area required will be the total number of squares times the area represented by one square. Squared tracing paper can be made by tracing or photocopying graph paper onto tracing paper. Alternatively the map itself may be photocopied onto squared paper instead of plain.

The average yearly streamflow (called the *runoff*) at points A and B can be approximated as follows:

Runoff = (Rainfall − Evaporation − Surface absorption) x Catchment area

Remember to convert the units of rainfall and evaporation from mm/yr to m/yr, and of the catchment area from km^2 to m^2, giving the runoff in m^3/yr.

The average daily flow (ADF) is therefore equal to runoff/365.

Evaporative loss

A great deal of water evaporates if the air is dry and the sky is clear. Wind increases the evaporation, as does the length of the day and air temperature.

A simple method of accounting for evaporative loss is sometimes possible due to the fact that the relationship between runoff and rainfall is often similar in different valleys in the same region of the country. If hydrological records for other nearby rivers are available, these will include measurements of rainfall compared to runoff, collected onto rainfall-runoff graphs, as shown in Figure 4.4. The straight line on the graph demonstrates that in this region the runoff is roughly a constant proportion of the rainfall. Therefore if you know the annual rainfall at a specific place within the region, you can find the runoff directly from the graph. Note that this method takes subsurface flow loss into account as well as evaporation losses.

Where rainfall-runoff data is not available, evaporative loss must be accounted for in some other way. Considerable work has been

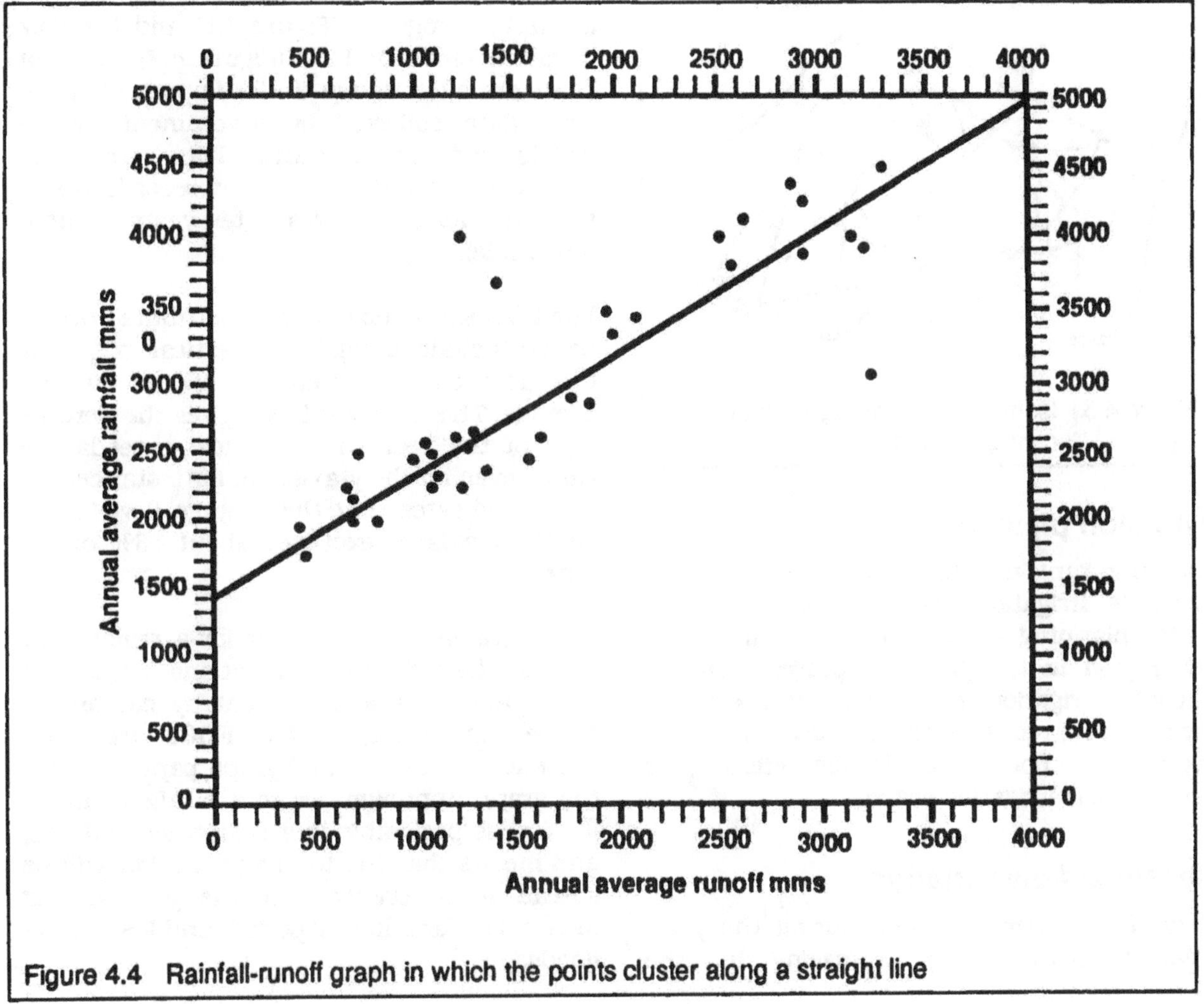

Figure 4.4 Rainfall-runoff graph in which the points cluster along a straight line

done to develop theoretical models to do this, a good example being Penman's Theory. This method and others are covered in E.M.Wilson's *Engineering Hydrology*. The reader is strongly recommended to consult this text before undertaking any but the simplest hydrological studies. A further source of information on methods in hydrology for micro-hydro catchments is the NRECA Foundation, whose address is in the bibliography.

Absence of local rainfall data

So far we have assumed that rain gauge records are available. If not, one or more of the following alternatives my be considered:

- if you have two or more years to wait for planning and finance clearance, immediately set up and monitor about three gauges in the region of interest. Do not use these records on their own, as two years worth of data can be misleading (ten years' data are ideally required) but correlate them with other national or regional data, such as isohyet maps (explained below);
- consult a professional hydrologist;
- use correlation methods, as given in texts on hydrology;
- if you have some years lead-in time, measure flow directly by building a solid weir structure across the river (see Section 4.4) and take regular measurements over as long a period as possible.

Isohyet maps

Often rainfall data are available in the form of *isohyet maps* as in Figure 4.5. These show lines of constant rainfall. They should never be used as a single indication of rainfall, but are sometimes useful as a check on other indications. On the whole their use should be avoided, since in micro-hydro applications the catchments are too small for isohyets to be sufficiently accurate.

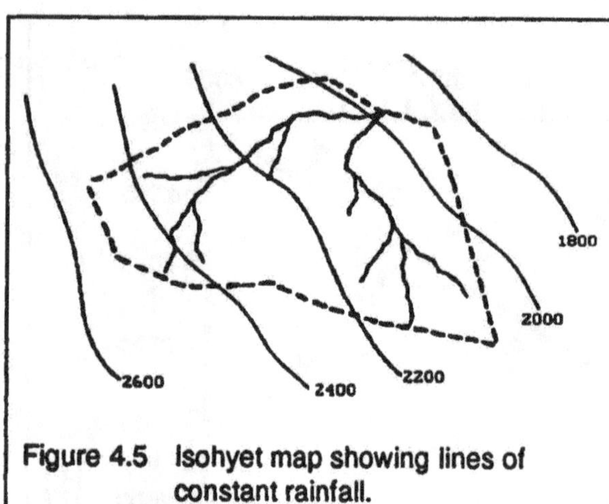

Figure 4.5 Isohyet map showing lines of constant rainfall.

Irrigation planning

An important use for water flow in many areas is irrigation. An essential aspect of hydro planning is to include the farmers of the region in the planning process, and to allow for irrigation requirements in the catchment being considered because they will reduce the flow available for micro-hydro power in the irrigation season.

Seasonal flow variation

River flow generally varies during the year. There are two ways of expressing this: the annual hydrograph (Figure 4.6) and the flow duration curve or FDC (Figure 4.7). Both of these are often drawn up for important rivers from data collected by government hydrologists over many years. They are most useful, especially if based on records taken every day for several years (ten years or more if possible).

The FDC shows how flow is distributed over a period (usually a year). The vertical axis gives the flow as a percentage of the annual average. The horizontal axis gives the percentage of the year that the flow exceeds the value given on the y-axis. So, for instance, the graph indicates that the average flow (100% on the y-axis) is exceeded about 33% of the time.

FDCs are often very similar for a region, but can be affected by soil conditions, vegetation cover, and to a lesser extent by catchment shape. As in Figure 4.7, FDCs are often presented on log/normal graph paper, so that the graph approximates to a straight line. A flat line is preferable to a steeply sloping one, and means that the total annual flow will be spread more evenly over the year, giving useful flow for a longer period, and less severe floods.

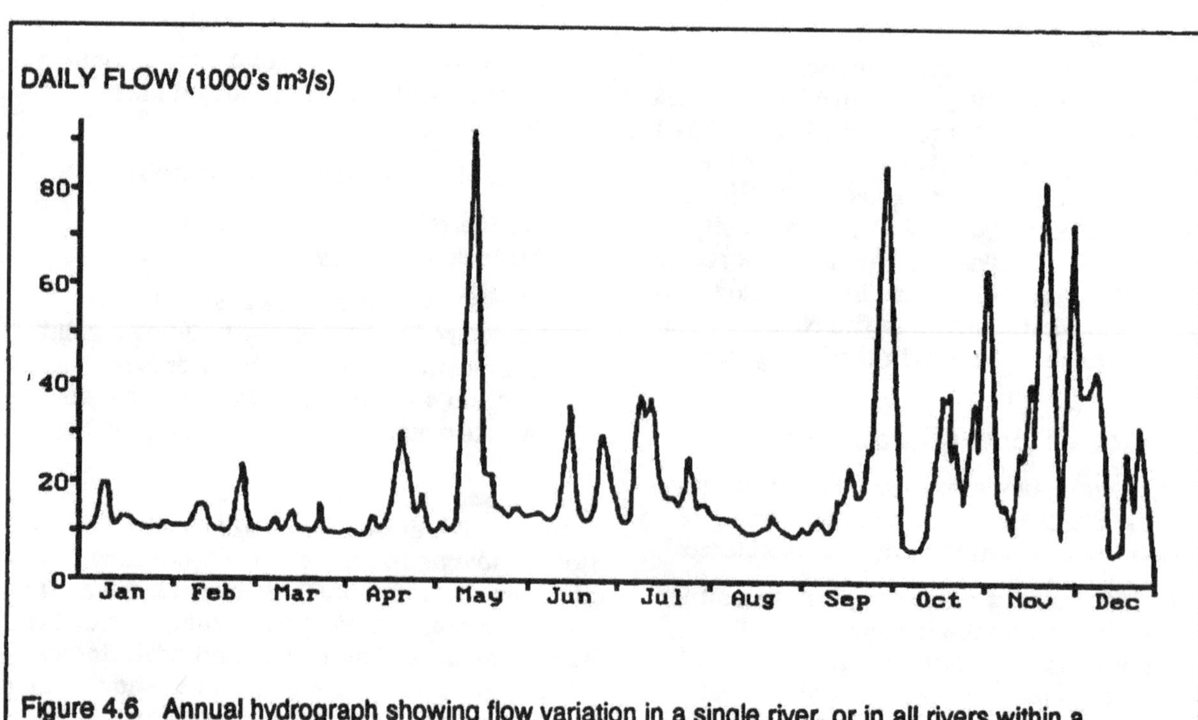

Figure 4.6 Annual hydrograph showing flow variation in a single river, or in all rivers within a particular region.

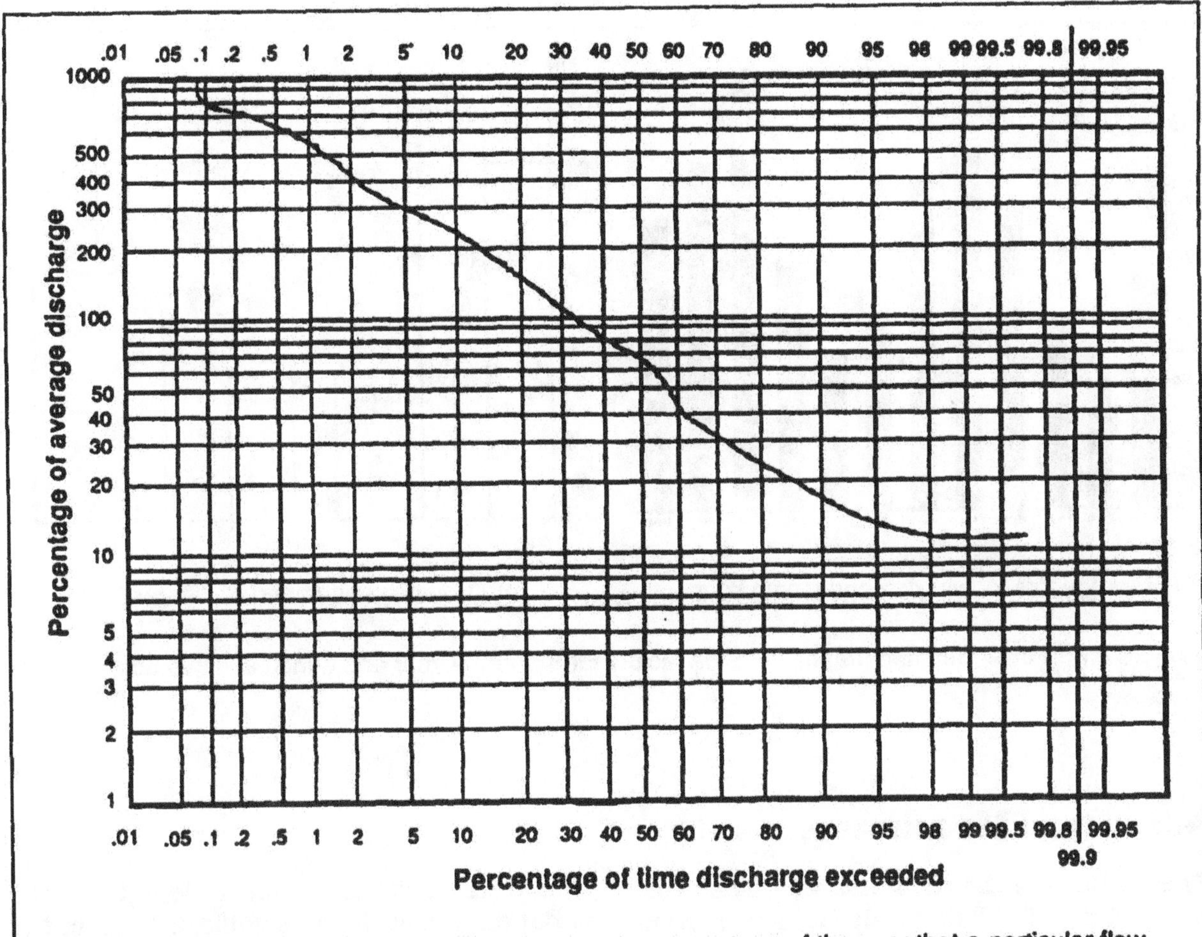

Figure 4.7 Flow Duration Curve (FDC) indicating the percentage of the year that a particular flow rate is exceeded.

Characteristics that give rise to a flat FDC are:

- deep soil
- heavy vegetation (e.g. jungle)
- long, gently sloping streams
- bogs, marshes
- even rainfall (temperate or two monsoons).

A steep flow duration curve is not good for micro-hydro as it implies a *flashy* catchment area i.e. one which is subject to extremes of flash floods and droughts. Factors which cause a catchment to be flashy are:

- rocky, shallow soil
- lack of vegetation cover
- steep, short streams
- uneven rainfall (frequent storms, long dry periods).

Matching supply and demand

The electrical power demand will vary both during the day and seasonally through the year. This is the case, for example, in a Sri Lankan tea factory, where an annual demand profile broken down by months might be as illustrated in Figure 4.8. A typical daily variation in load is also shown; clearly there are large variations in demand at different times of the day.

If the hydro scheme is at times producing more power than is needed, then some sort of governor or load controller will be required to ensure that the input power remains balanced with the generator output.

If the demand ever exceeds the supply, then arrangements must be made to manage the load differently so that the demand is evened out to avoid peaks. The alternative is to employ an auxiliary power source to take over some of the load during peak demand.

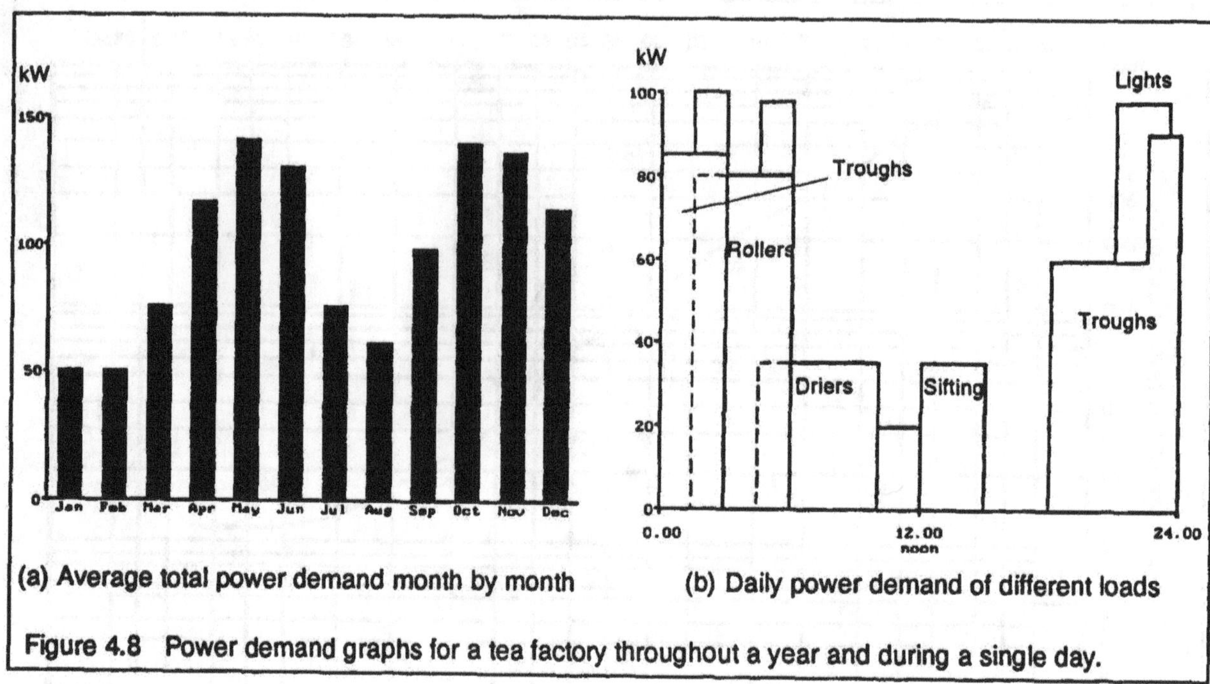

(a) Average total power demand month by month

(b) Daily power demand of different loads

Figure 4.8 Power demand graphs for a tea factory throughout a year and during a single day.

4.3 Head Measurements

The meaning of the term *head* was illustrated in Figure 2.3. The total (or *gross*) head available is one of the most important details required from a site survey. During pre-feasibility studies, various alternative sites may be examined, and a fast and approximate method adopted for finding the head. At the design and specification stage, more accurate techniques may be needed. This section looks at the various methods available.

Some measurement methods are more suitable on low-head sites, but are too tedious and inaccurate on high-heads; some are only suitable on high-head sites, being insufficiently accurate at low-heads. If possible, it is wise to take several separate measurements of head at each site. Always plan for enough time to allow on-site comparison of survey results. It is best not to leave the site before analysing the results, as any possible anomalies or mistakes will be easier to check on site.

A further very important factor to be aware of is that the gross head is not strictly a constant but varies with the river flow. As the river fills up, the tailwater level very often rises faster than the headwater level, thus reducing the total head available. Although this head variation is much less than the variation in flow, it can significantly affect the power available, especially in low-head schemes where every half metre is vital. To assess the available gross head accurately, headwater and tailwater levels need to be measured for the full range of river flows.

Dumpy levels and theodolites

The use of a Dumpy level (or builder's level) is the conventional method for measuring head and should be used wherever time and funds allow. Such equipment should be used by experienced operators who are capable of checking the calibration of the device.

Dumpy levels are used with staffs to measure head in a series of stages, as illustrated in Figure 4.9. A Dumpy level is a device which allows the operator to take a sight on a staff held by a colleague, knowing that the line of sight is exactly horizontal. Stages are usually limited by the length of the staff to a height change of no more than 3m. A clear unobstructed view is needed, so wooded sites can be frustrating with this method.

Dumpy levels only allow a horizontal sight, but *theodolites* can also measure vertical and horizontal angles, giving greater versatility and allowing faster work.

It is best to use professional, accurately calibrated surveying equipment in all situations where a high level of accuracy is needed (especially on low-head sites) or where long open slopes are being surveyed. Such instruments cost between US$400 and US$3000 but can often be hired or borrowed. The most modern equipment incorporates laser measurements and automatic electronic read-outs, making for faster and more accurate readings, but it is obviously much more expensive and can need specially trained operators.

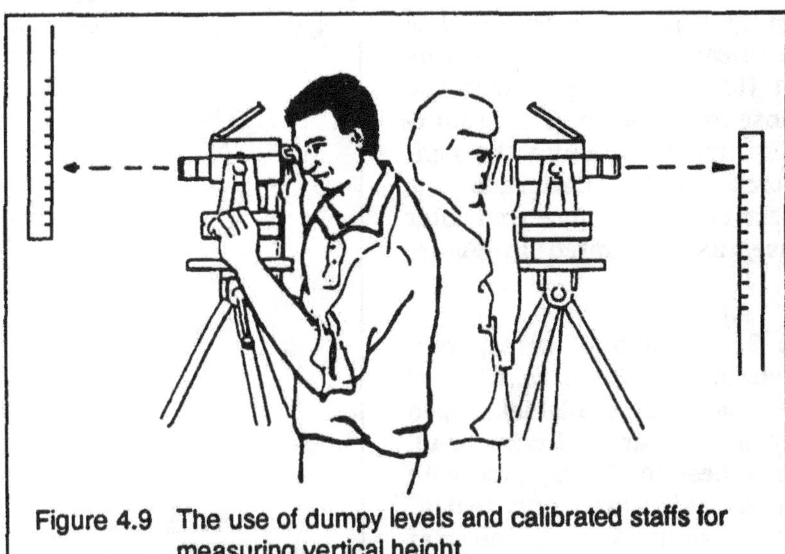

Figure 4.9 The use of dumpy levels and calibrated staffs for measuring vertical height

Sighting meters

Hand-held sighting meters (Figure 4.10) measure angle of inclination of a slope (they are often called inclinometers or Abney levels). They can be accurate if used by an experienced person, but it is easy to make mistakes and double checking is recommended. They are small and compact, and sometimes include range finders which save the trouble of measuring linear distance.

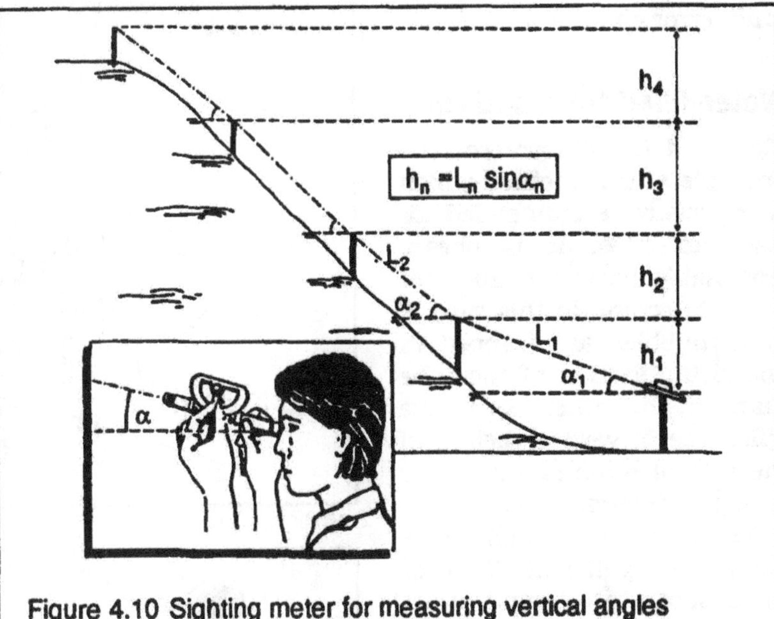

Figure 4.10 Sighting meter for measuring vertical angles

Since this method requires the linear distance along the slope to be recorded, it can have the advantage of doubling as a measure of the length of penstock pipe needed. These devices typically cost US$150-500 and are useful for any unwooded site. The error will depend on the skill of the user and will typically be between 2% and 10%.

Water-filled tube and pressure gauge

Shown in Figure 4.11, this is probably the best of the simple methods available, but it does have its pitfalls. The two sources of error which must be avoided, are out-of-calibration gauges and air bubbles in the hose. To avoid the first error, you should recalibrate the gauge both before and after each major site survey. To avoid the second, you should use a clear plastic tube allowing you to see bubbles. A 5cm (vertical height) air bubble seen in a rising section of the tube will give a gauge reading which is 5cm too low: add 5cm to obtain the correct estimate of head. Bubbles trapped in coils of tube on the ground are not important, nor are bubbles trapped at the top of vertical bends.

This method can be used on high-heads as well as low ones, but the choice of pressure gauge depends on the head to be measured. An added bonus of this technique is that the hose can be used as a measuring tape to measure the penstock length. Calibration is best checked before and after use, as illustrated in Figure 4.11

A 20m length of transparent plastic pipe (8mm vehicle fuel pipe is widely available) is good for sites of around 60m head. It is best to fill the pipe with water beforehand and seal it, and check for air bubbles before use. Equipment cost is around US$50 and the maximum error about 2%.

Figure 4.11 Water-filled tube and pressure gauge

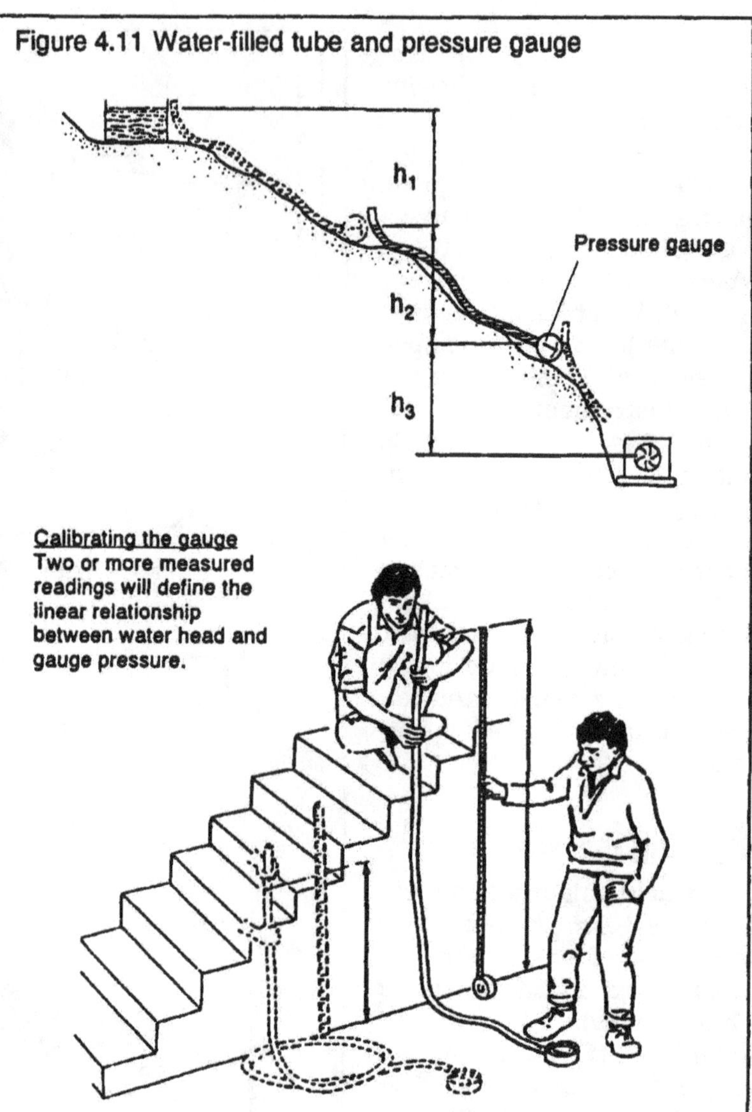

Water-filled tube and rod

Figure 4.12 illustrates the principle of this method which is especially recommended for low-head sites. It is cheap, reasonably accurate, and not prone to errors. In this case, if more bubbles are trapped in one rising section of the tube than in the other, then the difference in vertical height of the sets of bubbles will cause an equal difference in the head being measured, though this is usually insignificant. Two or three separate attempts must be made to ensure that your final results are consistent and reliable. In addition the results can be cross-checked against measurements made by another method, for instance by water-filled hose and pressure gauge. The accuracy of this method can be quite good even when a person (rather than a ruler) is used as the reference height. This method has been taught to village groups and farmers in self-help hydro projects in developing countries so that they can survey their own sites. Equipment cost is around US$5.

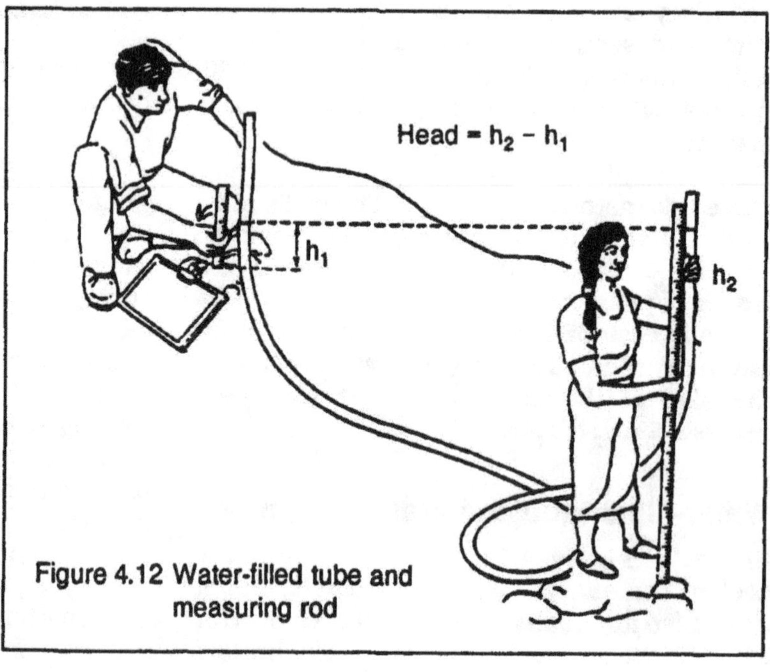

Figure 4.12 Water-filled tube and measuring rod

Spirit level and plank

This method is identical in principle to the water-filled tube and rod method. The difference is that a horizontal sighting is established not by water levels but by a carpenter's spirit level placed on a reliably straight plank of wood, as in Figure 4.13. On gentle slopes the method is very slow, but on steep slopes it is useful. Mark one end of plank and turn it at each reading to cancel the errors. The error is around 2%, the cost of equipment about US$10 to $20. This is only really suitable for low-head sites.

Maps

Large-scale maps are very useful for approximate head values, but are not always available or totally reliable. For high-head sites (>100m) 1:50 000 maps become useful and are almost always available.

Altimeters

These can be useful for high-head pre-feasibility studies. Surveying altimeters in experienced hands will give errors of as little as 3% in 100m. Atmospheric pressure variations need to be allowed for, however, and this method cannot be generally recommended except for approximate readings. Surveying altimeters cost around US$700.

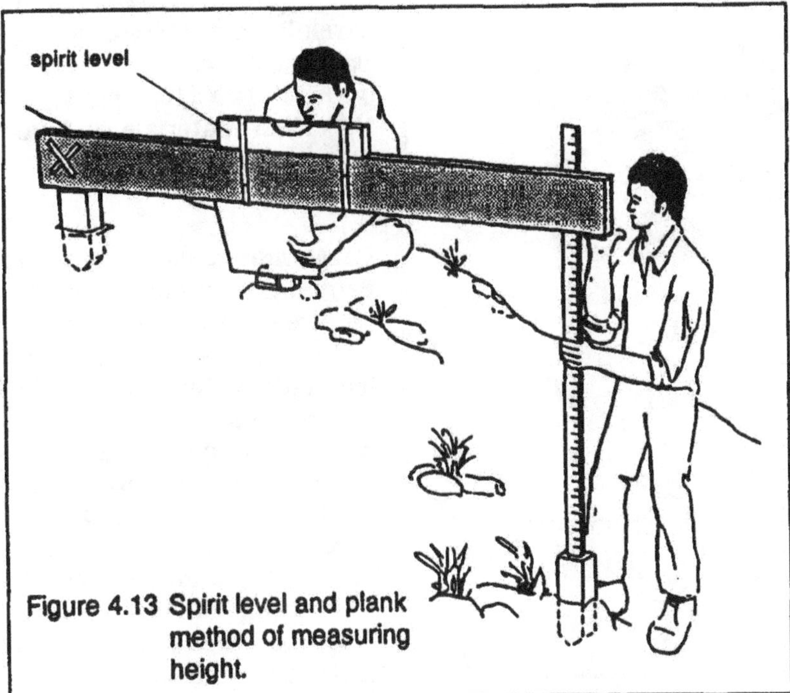

Figure 4.13 Spirit level and plank method of measuring height.

4.4 On-Site Measurement of Flow

The purpose of the hydrology study is to predict the variation in the flow during the year. Since the flow varies from day to day, a one-off measurement is of limited use. In the absence of any hydrological analysis, a long-term measuring system may be set up. Such a system is often used to reinforce the hydrological approach and is also the most reliable way of determining actual flow at a site. One-off measurements are useful to give a spot check on hydrological predictions.

The flow measuring techniques described here are:

- the weir method
- stage control method
- the salt gulp method
- the bucket method
- the float method
- current meters

It is necessary to study the distinctive features of each of these in order to find a suitable method for any particular site. Most of the literature on flow measurement refers to rivers, not small, turbulent streams. Different techniques are often needed for the streams typical in much micro-hydro work.

Measuring weirs

A flow measurement weir has a notch in it through which all the water in the stream flows. We will concentrate here on a rectangular notch, useful typically for flows in the region of 1000l/s. The flow rate can be determined from a single reading of the difference in height between the upstream water level and the bottom of the notch (see Figure 4.14). For reliable results, the crest of the weir must be kept sharp and sediment must be prevented from accumulating behind the weir. Sharp and durable crests are normally formed from sheet metal, preferably brass or stainless steel, as these do not corrode.

The formula for a rectangular notched weir is:

$$Q = 0.67\ C_w \sqrt{(2g)}\ (L - 0.2h)\ h^{1.5}$$

where
Q = flow rate (m^3/s)
C_w = the coefficient of discharge
L = the notch width (m)
h = the head difference (m) (see Figure 4.14)
g = acceleration due to gravity ($9.81 m/s^2$)

If C_w is taken, typically, as 0.6, then the equation becomes:

$$Q = 1.8\ (L - 0.2h)\ h^{1.5}$$

For instance, if the notched weir has width 2.2m, and the upstream water level is 34cm above the weir crest, then the flow rate is

$$Q = 1.8 \times (2.2 - 0.2 \times 0.34) \times 0.34^{1.5}$$

$$= \underline{0.761}\ m^3/s \quad (\text{or } 761 l/s)$$

Figure 4.14 Rectangular notch measuring weir

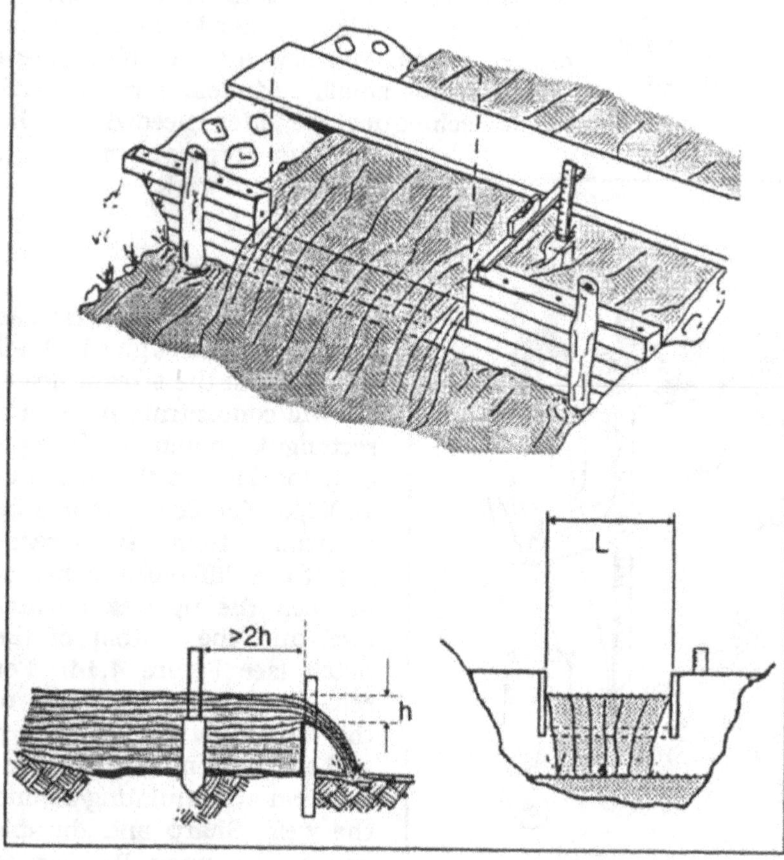

Weirs can be timber, concrete or metal and must always be oriented at right angles to the stream flow. Siting of the weir should be at a point where the stream is straight and free from eddies. Upstream, the distance between the point of measurement and the crest of the weir should be at least twice the maximum head to be measured. There should be no obstructions to flow near the notch, and the weir must be perfectly sealed against leakage.

The crest of the weir should be high enough to allow water to fall freely leaving an air space under the outflowing sheet of water. The surface speed behind the weir must be less than 0.15m/s.

Temporary measuring weirs
These are used for short-term or dry-season measurements and are usually constructed from wood and staked into the bank and stream bed. Sealing problems may be solved by attaching a large sheet of plastic and laying it upstream of the weir held down with gravel or rocks. It is necessary to estimate the range of flows to be measured before designing the weir, to ensure that the chosen size of notch will be correct.

Even temporary weirs require considerable effort to design, construct and install, and other methods should be carefully considered. A temporary weir for a flow range of 100 to 1000l/s might cost US$50 in materials and take four person-days to construct and install.

Permanent measuring weirs
Permanent measuring weirs require the same construction techniques as diversion weirs (covered in Section 5.4). The use of permanent weirs may be a useful approach for small streams, but larger streams might better be measured by staging (explained below). A permanent weir for a flow range of 100 to 1000l/s might cost US$1000 to $2000 to design and install.

Stage-discharge method

Once set up, this method provides an instant measurement of the flow at any time. It depends on a fixed relationship between the water level and the flow at a particular section of the stream. This section (the *contour* section) is calibrated by taking readings of water level and flow (*stage* and *discharge*) for a few different water levels, covering the range of flows of interest, so as to build up a stage-discharge curve. During calibration the flow does not have to be measured at the contour section itself. Readings can be taken either upstream or downstream using, for instance, a temporary weir, as long as no water enters or leaves the stream in between. The stage-discharge curve should be updated each year. A calibrated staff is then fixed in the stream (Figure 4.15) and the water level indicated corresponds to a river flow rate which can be read off the stage-discharge curve.

The constraints on the contour section are:

- it should be in a straight section of the stream;
- the flow should be in a single channel at all times;
- the stream bed should not be subject to scour or growth of weeds;
- the banks should be permanent and high enough to contain floods;
- access to read the measuring staff should be good.

Figure 4.15 Stage-discharge method: staff fixed at the contour section

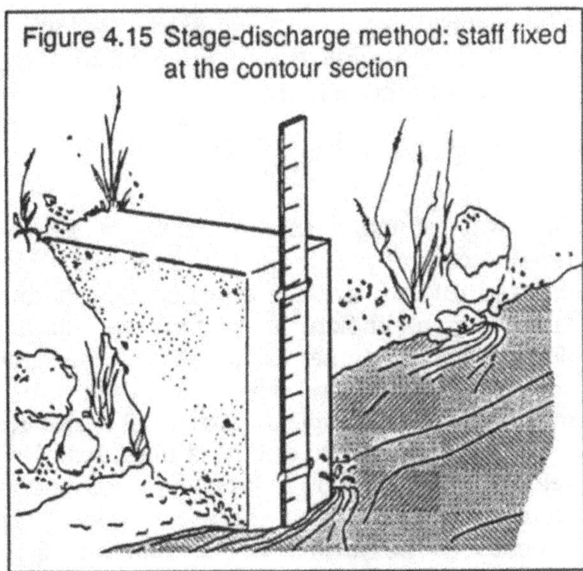

The calibrated staff is usually of wood and must be securely fixed. It can be placed in a sheltered part of the stream to avoid damage, usually at one side. A second reference (e.g. a mark on a large rock) should be made and levelled with a mark on the staff. This can be used to check that the staff has not been disturbed.

Useful references for stage discharge work are:

WMO Operational Hydrology Report no.13 (WMO, no.519).

Manual on Stream Gauging, Vol.1: Fieldwork.

Herschy,R.W. (1985), *Streamflow Measurements*,(Elsevier).

WMO Guide to Hydrological Practices, Vol.1 (WMO, no.168).

'Salt gulp' method

The 'salt gulp' method of flow measurement (Figure 4.16) is adapted from dilution gauging methods with radioactive tracers used for rivers. It has proved easy to accomplish, reasonably accurate (error <7%), and reliable in a wide range of stream types. It gives better results the more turbulent the stream. Using this approach, a spot check of stream flow can be taken in less than 10 minutes with very little equipment.

A bucket of heavily salted water is poured into the stream. The cloud of salty water in the stream starts to spread out while travelling downstream. At a certain point downstream it will have filled the width of the stream. The cloud will have a leading part which is weak in salt, a middle part which is strong in salt, and a lagging part which is weak again. The saltiness (salinity) of the water can be measured with an electrical conductivity meter. If the stream is small, it will not dilute the salt very much, so the electrical conductivity of the cloud (which is greater the saltier the water) will be high. Therefore low flows are indicated by high conductivity and vice versa. The flow rate is therefore inversely proportional to the degree of conductivity of the cloud.

The above argument assumes that the cloud passes the probe in the same time in each case. But the slower the flow, the longer the cloud takes to pass the probe. Thus flow is also inversely proportional to the cloud-pass-

Figure 4.16 The salt gulp method

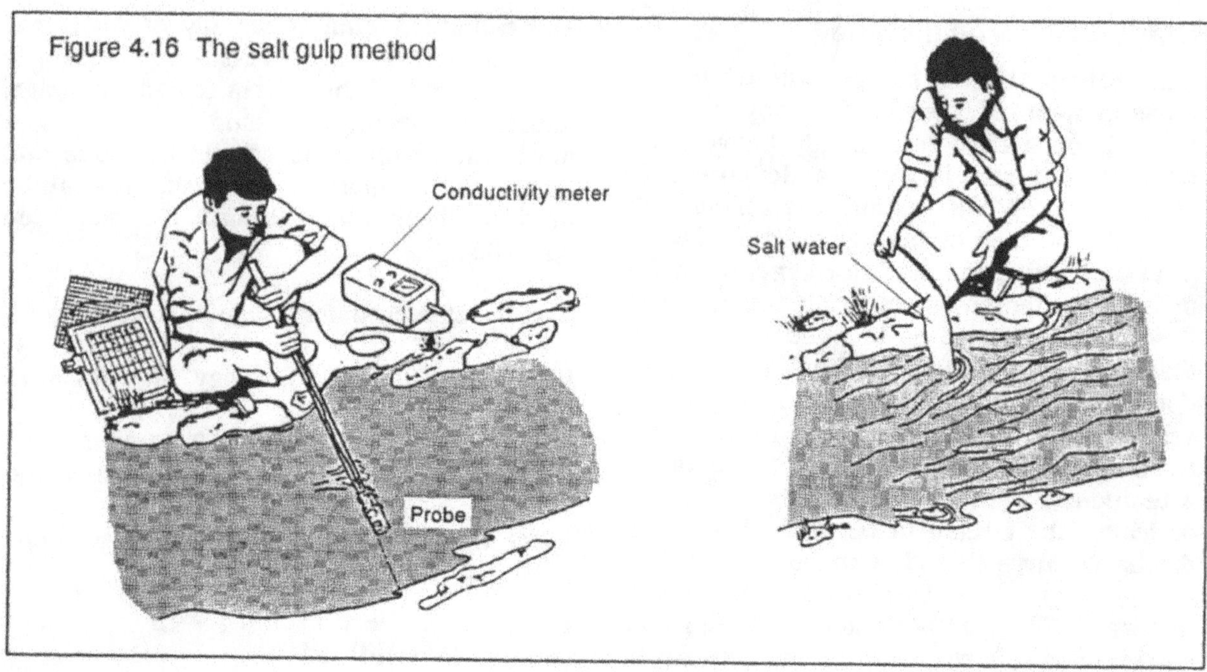

ing time. We will not attempt to cover the detailed mathematics here because the conductivity metre is usually supplied with detailed instructions.

The equipment needed for 'salt gulp' flow measurements is:

- a bucket
- pure table salt
- a thermometer (Range 0°–40°C)
- a conductivity meter (Range 0–1000 mS)
- an electronic integrator (optional).

Many conductivity meters are temperature corrected in which case a separate thermometer is unnecessary. An integrator (about US$200) will spare you the trouble of taking 5 second readings and plotting them out, but it can introduce errors. A useful guide to salt quantity is 100g for each $0.1 m^3/s$ of expected streamflow.

For a detailed description of the procedure and calculations, we refer you to the paper "Stream Flow Measurement by Salt Dilution Gauging" by A.Brown, ITIS, November 1983.

Bucket method

The bucket method is a simple way of measuring flow in very small streams. The entire flow is diverted into a bucket or barrel and the time for the container to fill is recorded. The flow rate is obtained simply by dividing the volume of the container by the filling time. Flows of up to $20 l/s$ can be measured using a 200-litre oil barrel.

The disadvantage of this method is that the whole flow must be channelled into the container. Usually a temporary dam has to be built and so the method is only practical for small streams, i.e. with flows in the range 10 to $20 l/s$. It can also only provide an occasional spot check of flow.

Float method

The principle of all velocity-area methods is that flow Q equals the mean velocity V_{mean} times cross-sectional area A:

$$Q = A \times V_{mean} \ (m^3/s)$$

One way of using this principle is for the cross-sectional profile of a stream bed to be charted and an average cross-section established for a known length of the stream (Figure 4.17). A series of floats, perhaps convenient pieces of wood, are then timed over a measured length of stream. Results are averaged and a flow velocity is obtained. This velocity must then be reduced by a correction factor which estimates the mean velocity as opposed to the surface velocity. By multiplying the average cross-sectional area by the

Figure 4.17 Charting the cross-sectional area of a stream

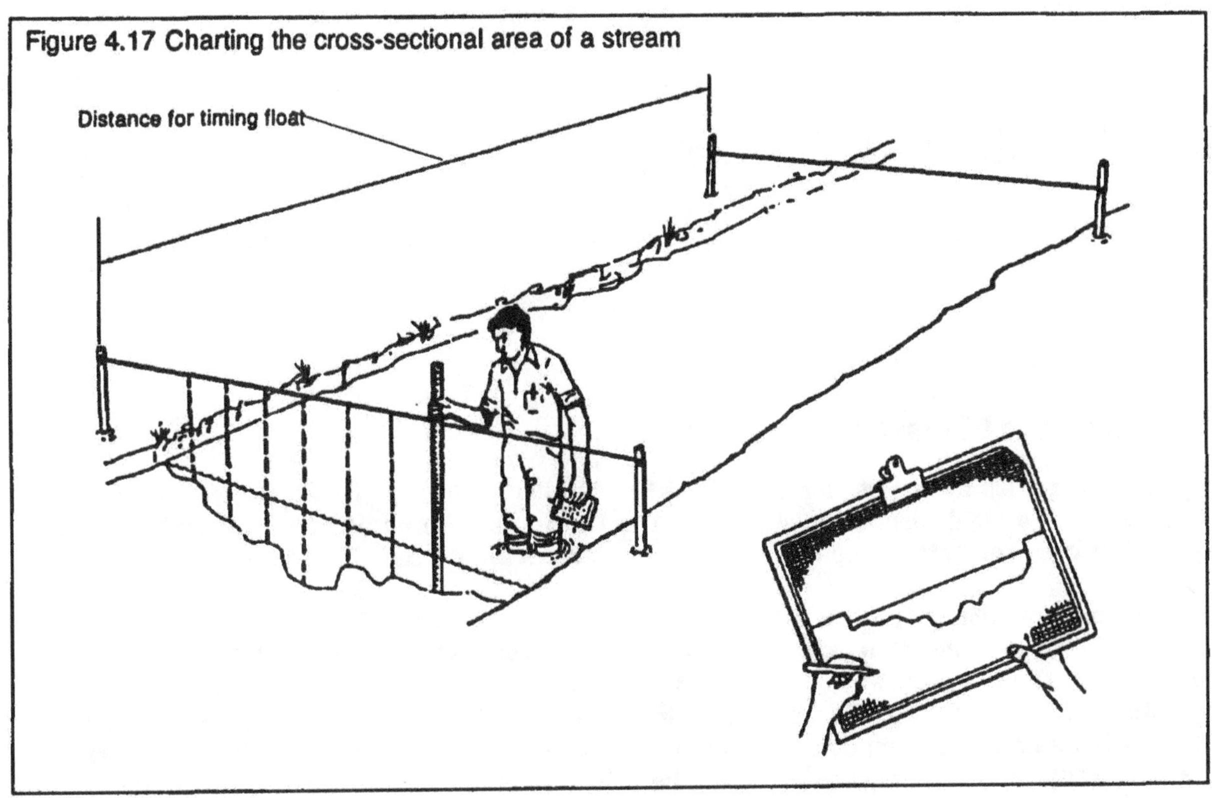

averaged and corrected flow velocity, the volume flow rate can be estimated.

Approximate correction factors to convert surface velocity to mean velocity are:

Concrete channel, rectangular, smooth	0.85
Large slow clear stream (>10m²)	0.75
Small slow clear stream (<10m²)	0.65
Shallow (<0.5m) turbulent stream	0.45
Very shallow (<0.2m) turbulent stream	0.25

Unless a smooth regular channel is considered, obtaining an accurate figure for the cross-sectional area of the stream will be very difficult and tedious. Therefore errors can be large, especially in shallow, rocky streams.

Current meters

These consist of a shaft with a propeller or revolving cups connected to the end. The propeller is free to rotate and the speed of rotation is related to the stream velocity. A simple mechanical counter records the number of revolutions of a propeller placed at a desired depth. By averaging readings taken evenly throughout the cross section, an average speed can be obtained which is more accurate than with the float method.

Current meters are supplied by their manufacturer with a formula relating rotational speed to the speed of the stream. Generally these devices are used to measure velocities from 0.2 to 5m/s with a probable error of approximately 3%.

The propeller must be submerged below the surface. Often the manufacturer will provide a marker on the propeller handle to indicate the depth of the blades. Accurate current meters are expensive (over US$1000) and, like the float method, the cross-section has to be measured as well, usually with low accuracy. Another source of error is turbulence. Any flow which is not parallel to the propeller shaft will not be measured correctly.

Civil Works 5

5.1 General Design Strategy

VARIOUS possibilities exist for the general layout of a hydro-scheme. Figure 5.1 shows four common arrangements.

The design approach should be first to work out the layout options that are technically feasible, and then to look at how they could be optimized for economy. For example, in Nepal, penstocks are very expensive compared to earth channels, so the penstocks tend to be very short and steep, and quite high head losses are accepted to keep costs down. In the UK, the correct design might use a longer, larger penstock and a shorter channel, reflecting the different costs of labour and capital-intensive components. In both cases certain design criteria must be met, for instance the flow velocity in the channel must be below the level that will cause erosion, and the pressure ranges in the penstock will have to be acceptable to the turbine.

For large-scale hydro, it is common to optimize the positions and sizes of all the major components. For micro-hydro it is more sensible simply to test the economic effect of reducing or increasing such basic factors as the penstock diameter and then to make judgements based on a sensible compromise between performance and cost.

There are a few essential points that must be considered during the early design of a hydro installation.

1. Use of the available head
Of utmost importance is the effect of the various design decisions on the net head delivered to the turbine. Components such as the channel and penstock cannot be perfectly efficient. Inefficiencies appear as losses of useful head or pressure. For example, a smaller diameter penstock is less efficient than a larger one because the greater friction causes an increased head loss.

2. Flow variation
The river flow varies through the year, but the hydro installation is designed to take a constant flow. If the channel overflows there will be serious damage to its surroundings. The weir and intake must therefore divert the correct flow volume whether the river is in low or high flow. The main function of the weir is to ensure that the channel flow is maintained when the river is low. The intake structure is designed to regulate the flow to within reasonable limits when the river is in high flow. Further regulation of the channel flow is provided by spillways.

3. Sediment
Flowing water in the river may carry small particles of hard and abrasive matter (sediment); these can cause wear to the turbine if they are not removed before the water enters the penstock. Sediment may also block the intake or cause the channel to clog up if adequate precautions are not taken.

4. Floods
Flood waters will carry larger suspended particles and will even cause large stones to roll along the stream bed. Unless careful design principles are applied, the diversion weir, the intake structure, and the embankment walls of the river may be damaged.

5. Turbulence
In all parts of the water supply line, including the weir, the intake and the channel, sudden alterations to the flow direction will create turbulence which erodes structures, and causes energy losses.

Figure 5.1 Possible layouts for run-of-the-river micro-hydro schemes

5.2 System Layout

The basic rule in laying out a system is to keep the penstock as short as possible. Penstock pipework is usually considerably more expensive than an open channel. Figure 5.2 illustrates the various components that are commonly needed in a micro-hydro installation.

Figure 5.3 reproduces an enlarged section of the map of Figure 4.2 showing scheme A as considered in Chapter 4. Three possible penstock routes are shown. The short penstock option will generally lead to the most economical scheme, but this is not always so. Considering each option in turn:

1. Short penstock

Here the penstock is short but the channel is long (see Figure 5.4a). The long channel is exposed to a greater risk of blockage, or of deterioration as a result of poor maintenance. Installing the channel across the steep slope shown may be difficult and expensive, or even impossible. The risk of the steep slope eroding may make the short penstock layout an unacceptable option, because the projected operation and maintenance cost of the scheme could be very great, and might outweigh the benefit of reduced initial cost.

2. Mid-length penstock

The longer penstock will cost more, but the expense of constructing a channel that can safely cross a steep slope may be saved (see Figure 5.4b). Even if the initial purchase and construction costs are greater, this option may be preferable if there are signs of instability in the steep slope. In some cases the soil may be particularly sandy and permeable and cause water to leak from the channel, in which case a shorter channel would be a wise option, to minimize water losses by seepage.

It is important to note that many of the problems associated with open earth channels, such as excessive seepage loss and blockage from falling debris, can be solved by the use of closed pipes, or by covering and lining the channels. Although the costs of improved channels are high, they will usually be lower than penstock costs.

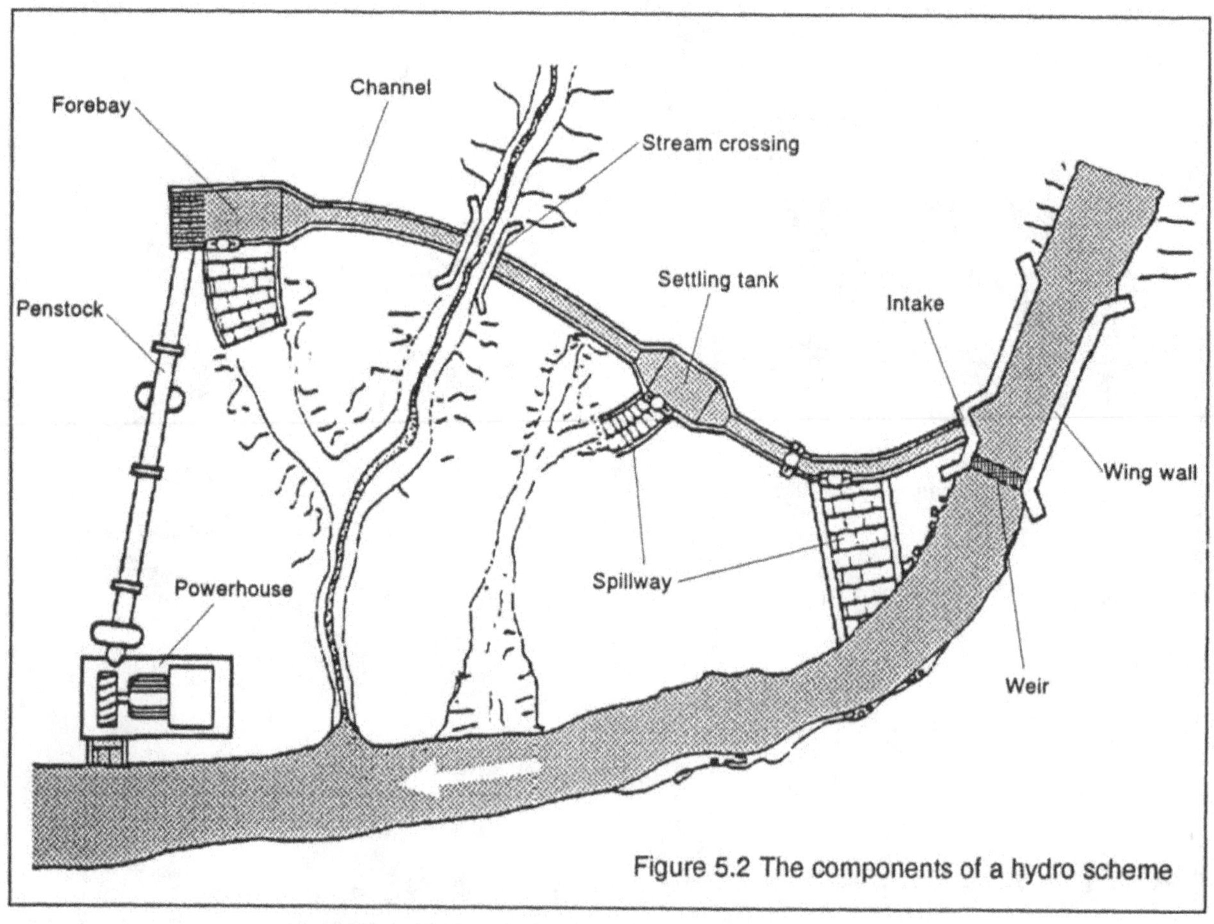

Figure 5.2 The components of a hydro scheme

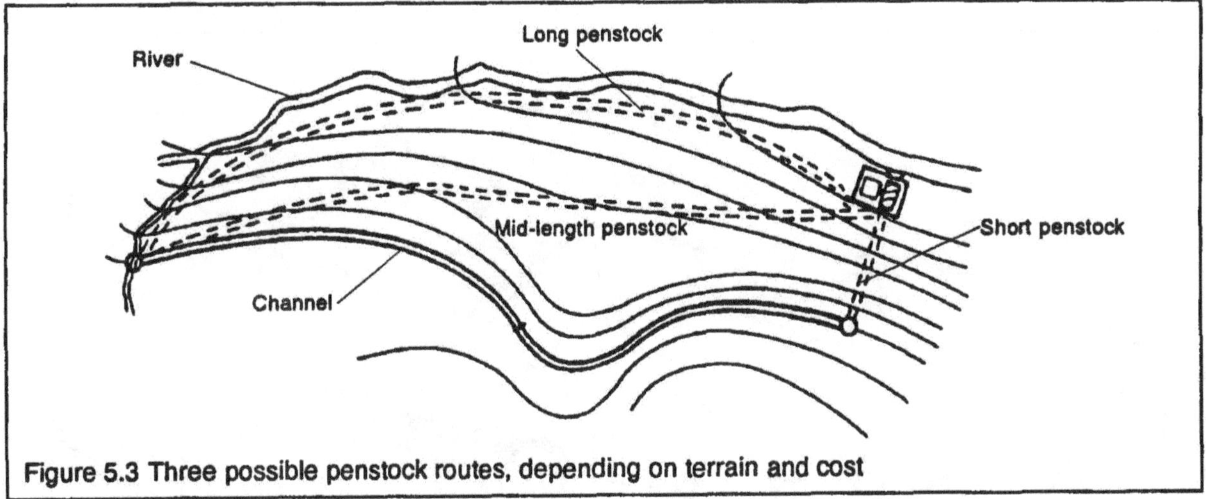

Figure 5.3 Three possible penstock routes, depending on terrain and cost

3. Long penstock
In this case the penstock follows the river, as shown in Figure 5.4c. If this layout is necessary because of difficult terrain on which to construct a channel, certain precautions must be taken. The most important one is to ensure that seasonal flooding of the river will not damage the penstock. It is always vital to calculate the most economic diameter of penstock; in the case of a long penstock the cost of errors in sizing will be especially high.

Cost

In order to make the best overall choice of system layout, some rapid costing process is necessary. Section 6.6 describes a method of estimating the penstock cost per unit length from the head, the maximum flow rate, and the chosen pipe material. This method must be applied to each of the options and combined with estimates of the associated channel costs.

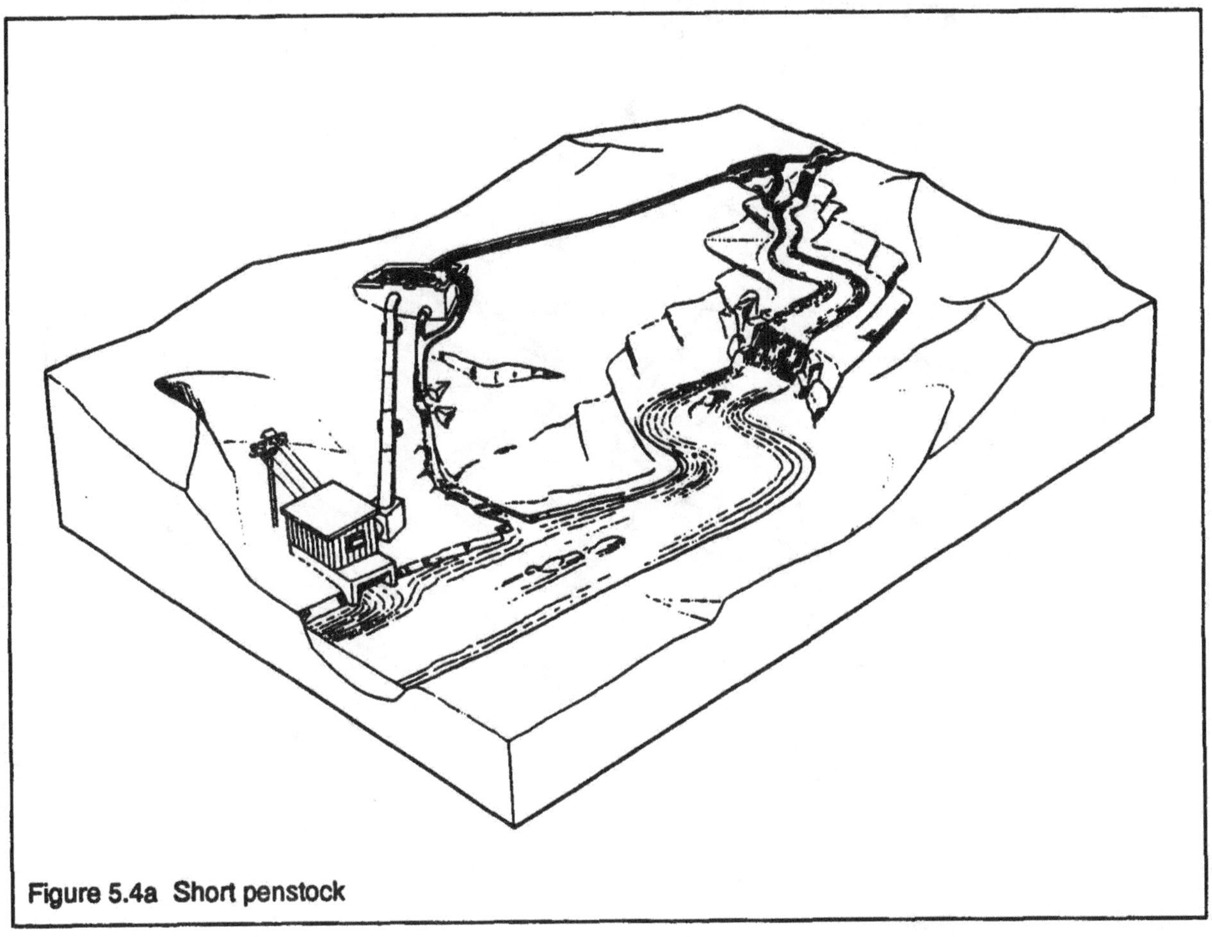
Figure 5.4a Short penstock

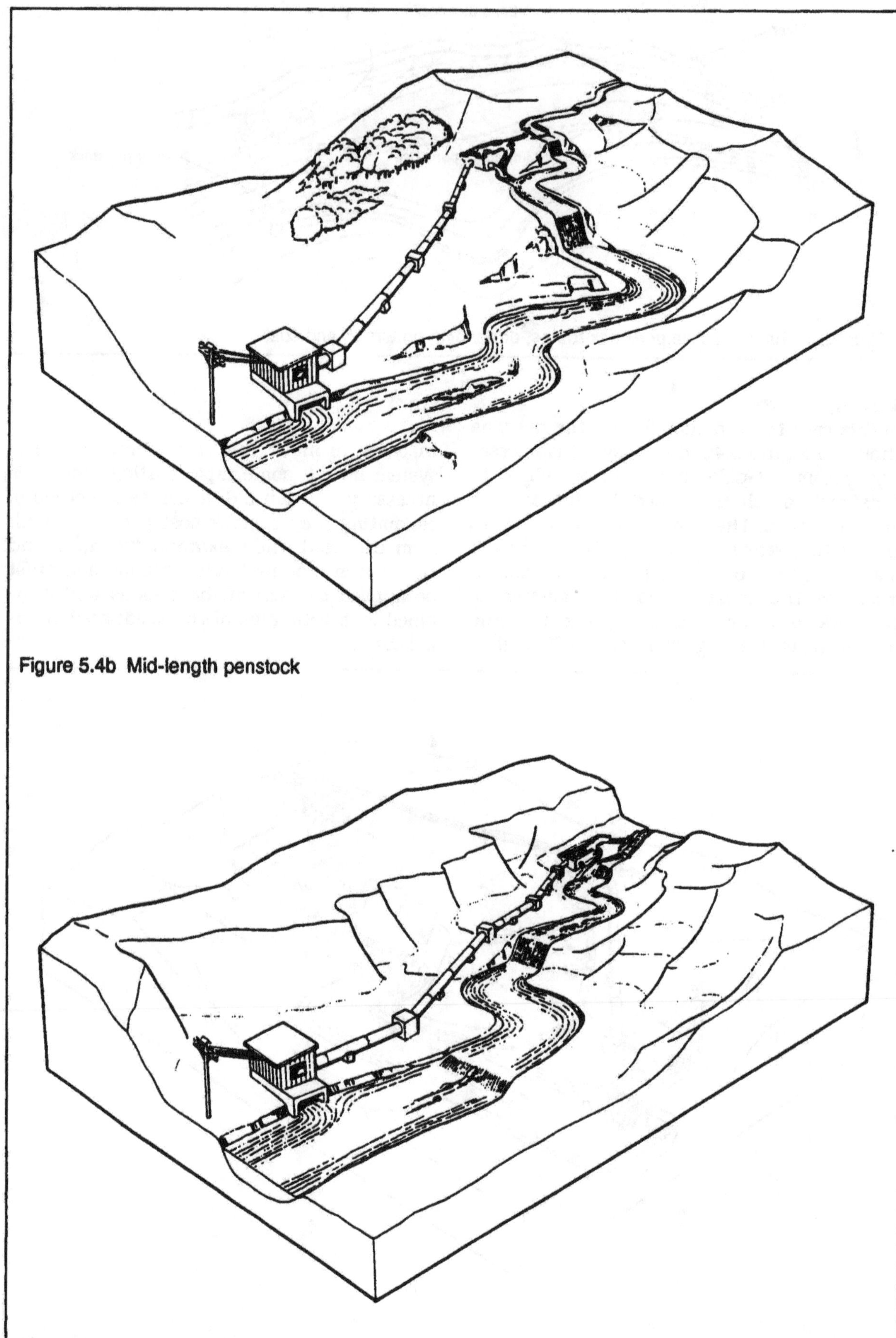

Figure 5.4b Mid-length penstock

Figure 5.4c Long penstock

5.3 Low-head Installations

Power is proportional to the product of flow and head, so for a given amount of power, a low-head installation needs proportionately more water than a high-head one. This means that for the same power, the conduits of water have to be bigger, but generally the channel and the penstocks can be short or even omitted.

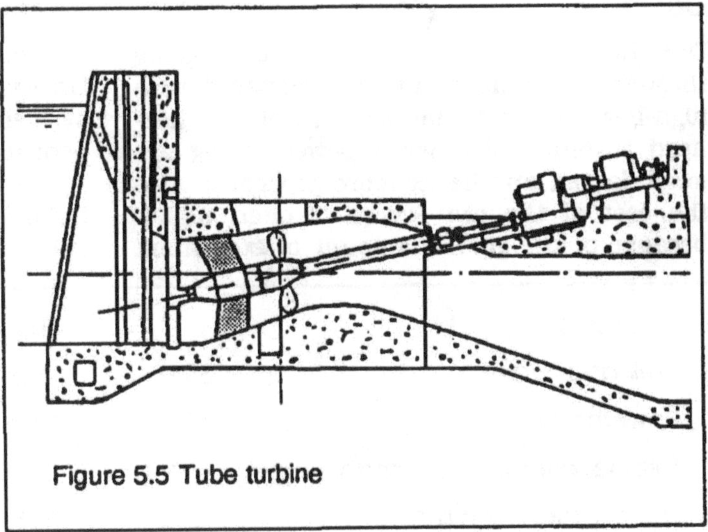
Figure 5.5 Tube turbine

Figure 5.5 shows a typical layout for a low-head 'tube' turbine with the tube and drive-shaft inclined. The configuration shown can run at heads of up to 10 or 15m. The turbine is mounted inside a tube, and this tube can be bolted to the penstock flange, or more often to an intake built directly into the weir. This layout allows easy access to the generator and permits the use of belt drives.

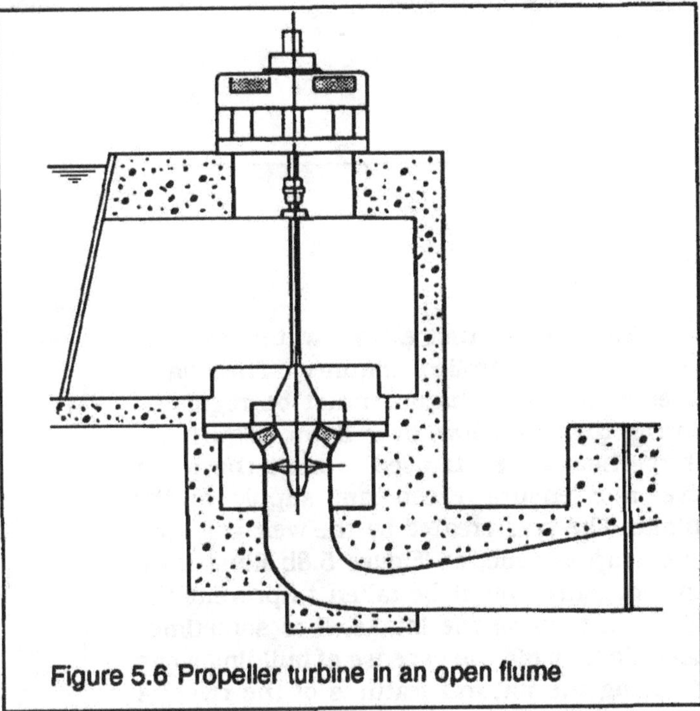
Figure 5.6 Propeller turbine in an open flume

The installation in Figure 5.6 is a vertical shaft propeller turbine in an open flume, which requires less civil works. A 'bulb' turbine, illustrated in Figure 5.7, offers the most compact installation, but for a micro-hydro plant bulb turbines are often expensive because they generally need an epicyclic (planetary) gearbox, and there is the complication of having the alternator sealed into a watertight enclosure in the centre of the stream of water.

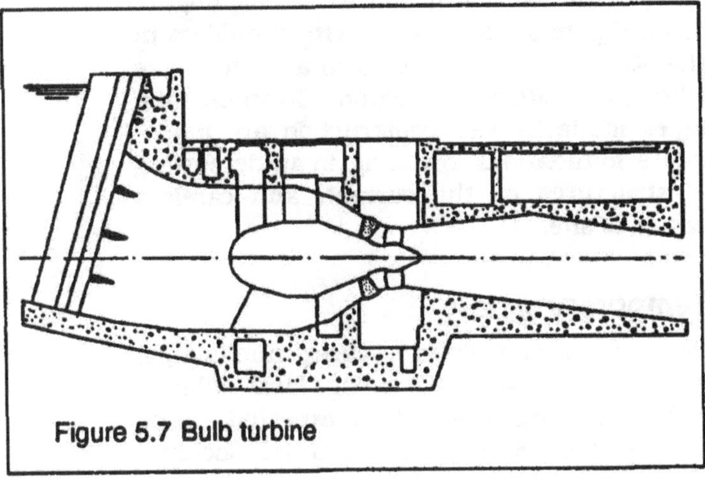
Figure 5.7 Bulb turbine

Low-head turbines can be used in parallel groups *(manifolding)*. For example, three identical 50kW units might be mounted alongside each other. This permits greater flexibility because, depending on flow availability and demand, one, two or all three turbines can be in use. It also allows standardization of turbines with potential manufacturing cost savings, particularly in the case of bulb turbines. Moreover, multi-turbine installations allow turbines to be overhauled or repaired without shutting down the entire system.

Costs

Due to the larger amount of water going through a low-head turbine compared with high-head or medium-head turbines, low-head turbines of a given power rating are much larger, and hence more expensive. Also the cost of the civil works is often higher because of the greater amount of excavation and concrete used in low-head installations. However, the cost of a penstock is much lower or even avoided which can considerably compensate for the other costs.

Below are examples of typical project costs for low-head and high-head schemes:

Low-head		High-head	
Weir and intake	35%	Intake, desilting and screening	12%
Powerhouse	15%	Penstock and installation	32%
Turbine, alternator & control system	35%	Powerhouse	9%
200m transmission line	5%	Turbine, alternator, controls	24%
Delivery and installation	5%	200m transmission line	5%
Sundries	5%	Delivery and installation	9%
	100%	Sundries	9%
			100%

5.4 Weirs

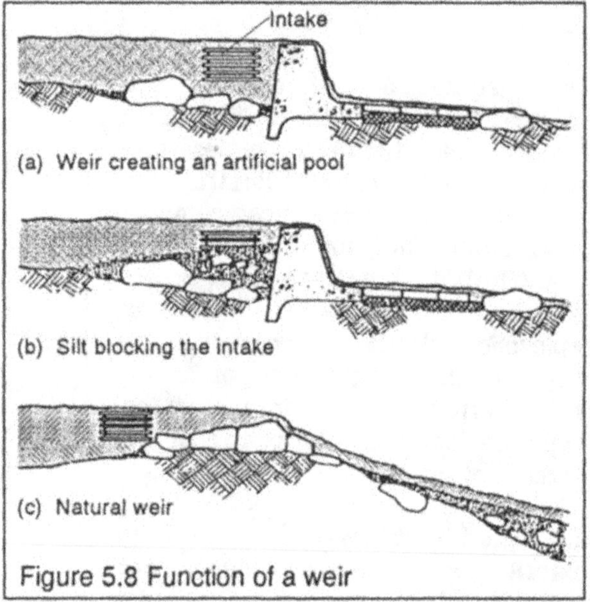

Figure 5.8 Function of a weir

A hydro scheme must extract water from the river in a controlled manner. The water diverted into the channel must be regulated during high and low river flows. Figure 5.8 shows how a weir is used to raise the water level and ensure a constant supply to the intake. The pool created by the weir will tend to silt up in time, as Figure 5.8b illustrates, and measures must be taken to prevent the silt from burying the intake. It is sometimes possible to avoid the expense of building weirs by using the natural features of the river. A natural permanent pool in the river may provide the same function as a weir (Figure 5.8c). Figure 5.9 shows how large boulders in the stream can be used to locate an intake with good natural protection. Common mistakes made in weir construction are inadequate foundations resulting in undercutting of structures by the current, and careless choice of site.

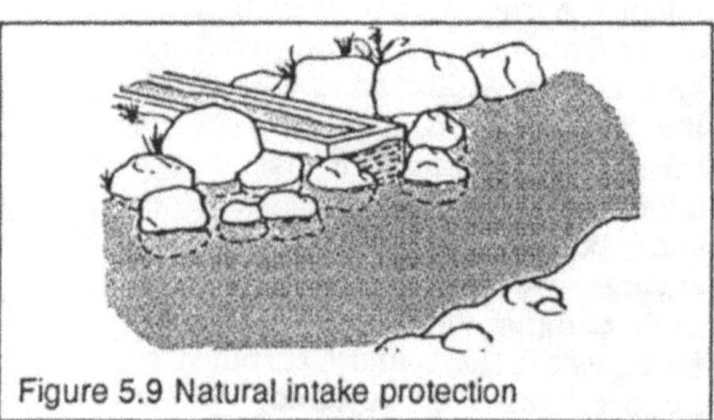
Figure 5.9 Natural intake protection

Temporary weirs

The temporary weir is a cheap and simple structure, constructed using local labour, skills, and materials. It is expected to be destroyed by annual or bi-annual flooding,

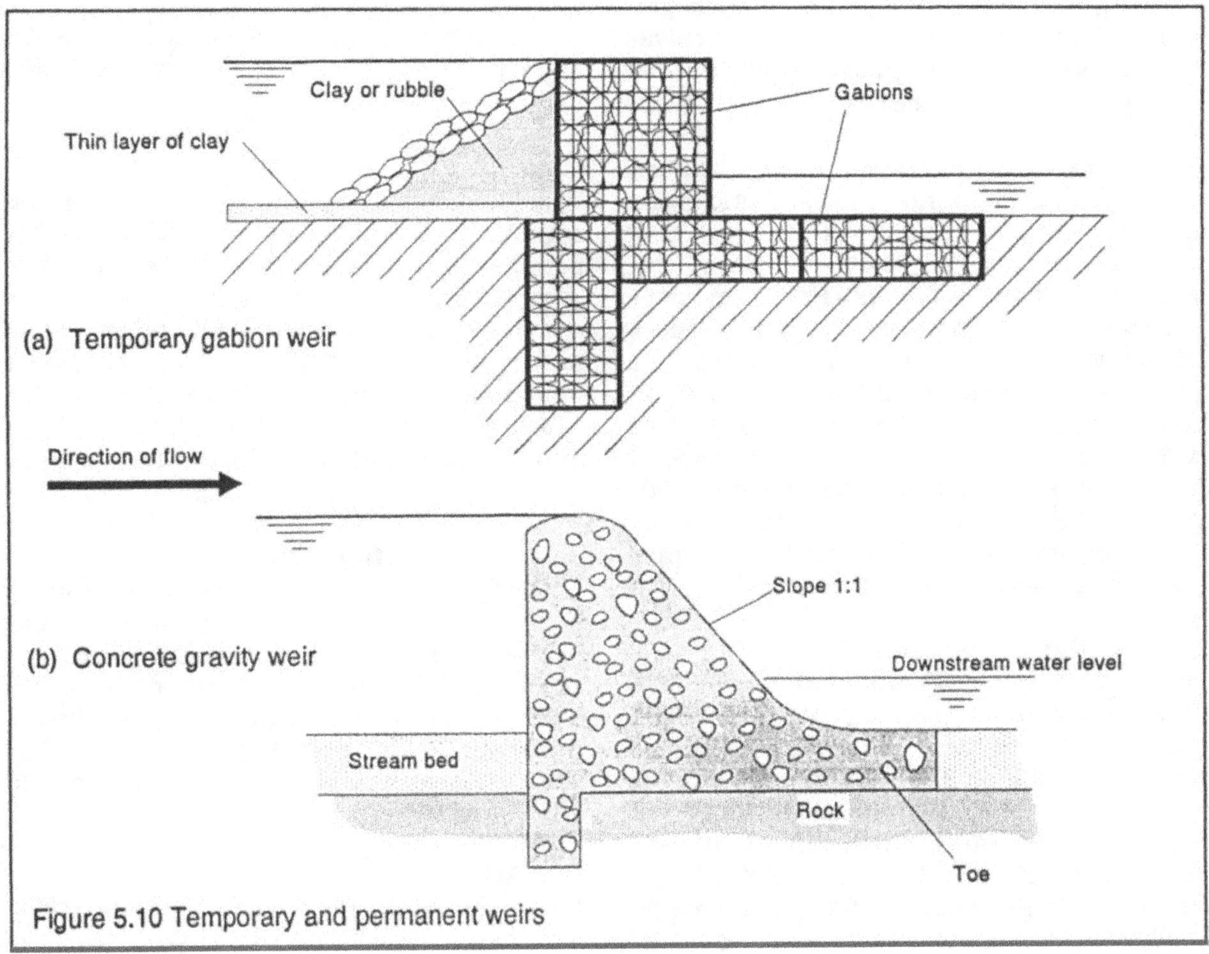

Figure 5.10 Temporary and permanent weirs

and rebuilding is planned in advance. Temporary weirs are commonly constructed with *gabions*, metal cages fabricated from galvanized steel wire and filled with rocks. The structure should be low, to minimize downstream damage if it is washed away by floods.

Sometimes traditional water management techniques are known to local people, and these may include methods of constructing temporary weirs. Detailed designs will not be covered here, but Figure 5.10a is one example of a gabion weir. A temporary weir may be only 5% of the cost of a permanent concrete weir, but it will require regular rebuilding.

Permanent weirs

The most important factor in selecting a satisfactory weir site is the presence of suitable foundations. These should consist of sound rock, free from fissures or cracks, and impervious to the action of water or air. Permanent weir construction is best done in concrete. Permanent earth weirs are difficult engineering structures and will not be dealt with here. Low concrete weirs which have base widths at least one-third greater than their height are automatically stable and are called *gravity weirs* (Figure 5.10b).

Before construction can begin, the stream has to be diverted to one side. Construction is then often done in two halves. A flushing pipe must be built into the low point of the weir to allow it to be drained for maintenance; this should be sealed on the downstream side. The most difficult part of construction is creating a seal across the stream bed. If the stream bed is continuous rock then there is little problem, but usually there is loose rock, earth and discontinuous bedrock to deal with. It is then necessary to excavate a narrow trench across the stream bed, down to the bedrock. This is filled with a cement grout (5 sand, 1 cement). The grout is driven down into cracks and gaps in the rock and forms a 'curtain' across the stream, preventing undercutting, which is the most common form of failure. This grouting can be done underwater, but inspection is best done in dry conditions (to gauge whether further excavation is needed), so pumps are usually needed to remove water from the site.

Sometimes the river has natural bedrock protrusions which form suitable foundations for the weir and result in a longer overall width of weir for the same amount of civil works (Figure 5.11). This extra width reduces the river level during a flood and so reduces the necessary heights of barriers such as the 'wing' walls which reinforce the bank on either side of the weir. The wing walls must be high enough to protect the banks in case of flood. This requires a calculation using 100-year flood data or a pessimistic estimate for maximum flow (e.g. perhaps up to 200 times the average daily flow ADF for tropical areas).

Curving the downstream face of the weir prevents suction which would, in time, cause damage to a face of rectangular section. The curve where the descending weir wall meets the downstream lip (the 'toe') prevents erosion of the weir foundation at this point (see Figure 5.10b). Steel reinforcement of the concrete is often used but is not always necessary if the quality of the concrete is good. It can limit impact damage on difficult, steep sites where large boulders may hit the weir during floods. Care must be taken in joining new concrete to old; large keystones must protrude from the old concrete. Steel helps in the keying process.

The cost of a weir depends largely on the expenditure on labour and the cost of access. It is normal to allow a generous budget to take account of bedrock problems and the effects of heavy rain during construction. A weir 20m long by 1.5m average height might cost anywhere between US$1,000 and US$20,000.

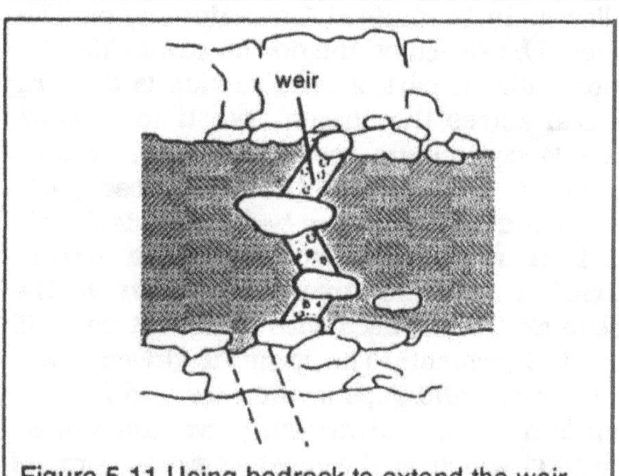

Figure 5.11 Using bedrock to extend the weir

Fish passes

Weirs across natural waterways will often be required to include some arrangement which permits fish to pass upstream. A good fish pass will (a) allow an easy passage for fish in a uniform flow of water, (b) ascend gradually and without any high barriers, (c) use a minimum of water, (d) have an entrance into which fish are easily directed, and (e) be durable and solid in construction. Three types of fish pass are often used:

Natural stream fishways

A small side-stream or spillway of gradual slope is used to connect the pond above the weir with the stream below. The disadvantages are that it usually means diverting an excessive flow of water, and the downstream entry/exit to the river has to be some distance from the weir, so fish will tend to miss it when swimming upstream.

Pool fishways

A series of pools are built to one side of the weir with a drop of 200 to 300mm between successive pool levels. Each pool has an exit orifice through which the water flows at a velocity of less then 2m/s. The pools can be constructed from concrete or bricks (or even timber) and should be large enough for the water flow within them to be low, so as to give ascending fish a chance to rest.

Fish ladders

These consist of a long and narrow inclined channel, with a slope of no more than about 1 in 10. Regular baffles or steps check the flow of water and provide pools of relatively still water as resting places for ascending fish. The baffles should be set perpendicular to the flow so that water striking them rises over their tops in a wave deep enough for fish to swim through when heading upstream.

Where river authorities have regulations regarding the provision of fish passes, it is important to find out the details of their preferred method for meeting the requirements. They will no doubt wish to approve the proposed hydro-installation both at the design stage and after completion.

5.5 Intake Design

A great many variations in intake and weir design are possible, to suit a wide range of natural conditions. To keep things simple, only a few of the possible options are discussed here.

Figure 5.12 illustrates two configurations of intake built into the wing wall: (a) a direct intake, and (b) a side intake. A direct intake diverts the flow more smoothly but a side intake is easier to construct. Large debris must not be allowed to enter the intake, nor to clog it up. With a side intake, the debris will be carried past by the flow, so a simple covering grill is sufficient (Figure 5.13). A direct intake requires an angled grill so that debris drawn towards the intake is diverted upwards.

Because the intake is sited in the river, it is often not easily accessible. During flood flow conditions, for instance, it may be dangerous to attempt any flow control operations near the intake mouth. Figure 5.14 is an example of a typical layout of intake and channel. After passing through the intake the water flows into a *headrace* (a fast-flow channel). It then continues past a spillway, through a silt basin, and on towards the forebay tank. The headrace allows control functions to be placed away from the river where access is safe. In many cases it is convenient to use a pipe rather than a channel for the headrace. Both must be able to withstand boulder damage in flood conditions.

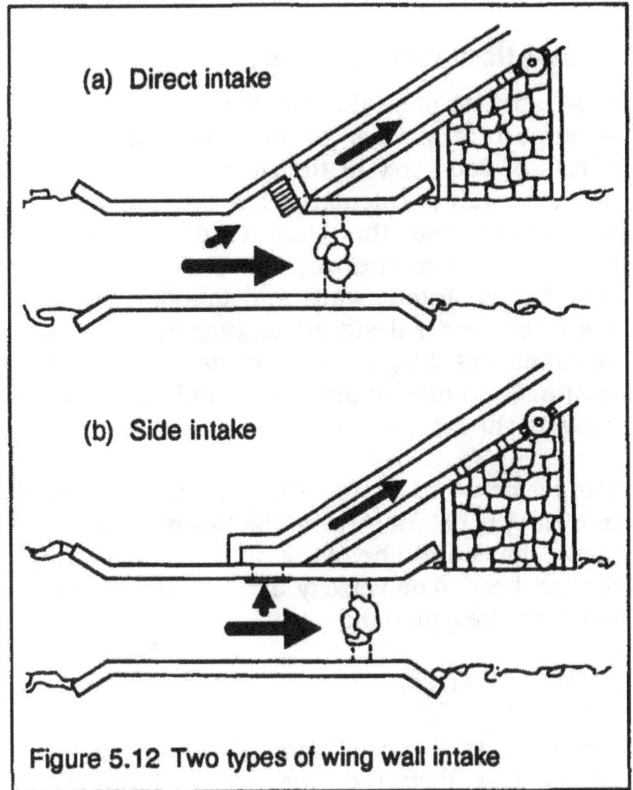

Figure 5.12 Two types of wing wall intake

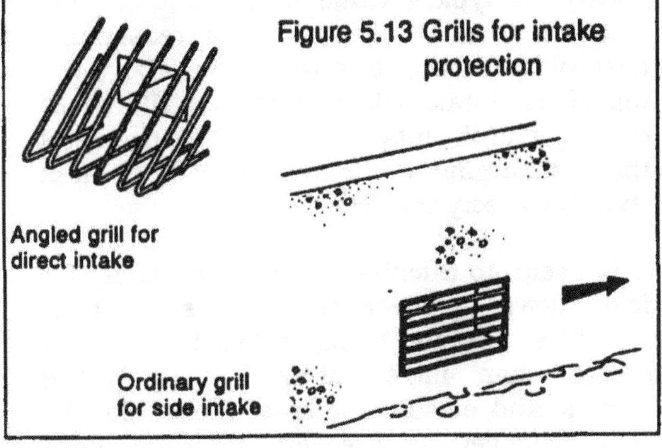

Figure 5.13 Grills for intake protection

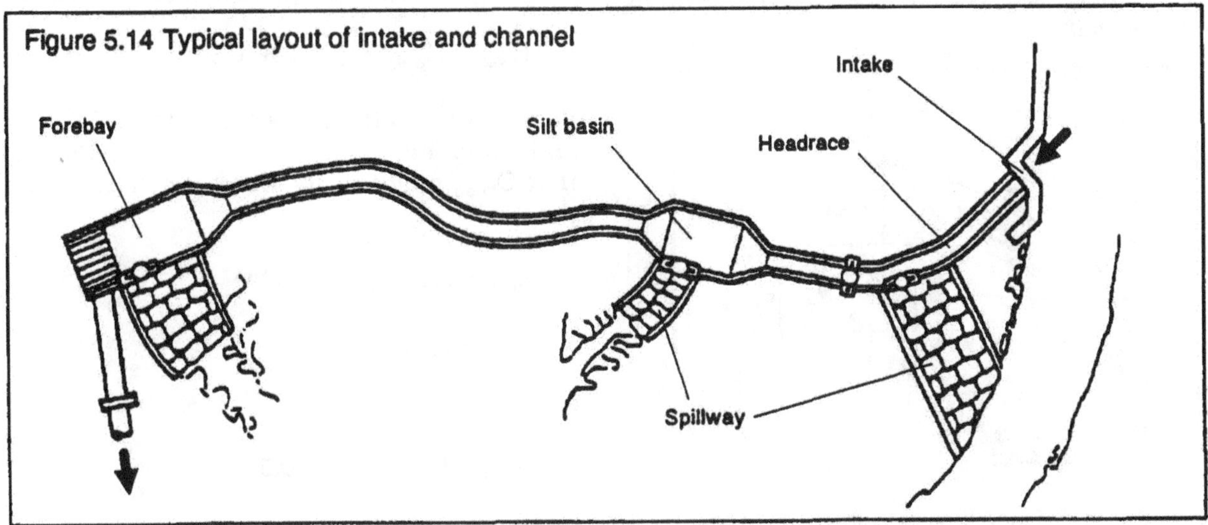

Figure 5.14 Typical layout of intake and channel

Intake flow calculations

In all aspects of intake and channel design, it is important always to bear in mind the effects of flood flow in the river. In particular it is essential to calculate what extra flow will be passed into the channel during flood conditions when the water level is considerably higher. Intake walls and intake mouths are often under-designed or even left out of schemes, resulting in frequent damage to the channel and high maintenance and downtime costs in the rainy season.

The velocity of water through the intake mouth V_i is determined by the height of water above the centre-line of the intake. This is the driving head. The velocity can be approximated from the equation:

$$V_i = C_d \sqrt{(2gh)}$$

where C_d is the coefficient of discharge (or energy loss factor) of the orifice, h is the driving head, and g is the acceleration due to gravity. A typical value of C_d is 0.6. This equation is only accurate for a reasonable head of water, e.g. for a water level above the top of the intake which is greater than half the depth of the intake orifice. Below this level the discharging water velocity will be less than the theory predicts.

It is useful to calculate the intake velocity for flood flows to check that excessively high velocities are not being induced; velocities greater than 4m/s will tend to erode the cement and masonry walls of the intake. To reduce the velocity, the weir length may need to be increased, perhaps leading to a re-siting of the weir.

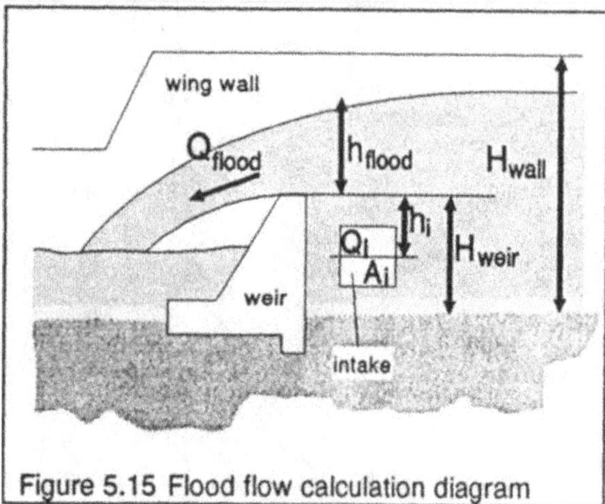

Figure 5.15 Flood flow calculation diagram

The flow of water through the intake Q_i, is simply the velocity of the water V_i multiplied by the area of the intake mouth A_i:

$$Q_i = A_i V_i$$

The intake must be designed to cope with the worst likely flood conditions. Important quantities to work out are:

- the height of wing wall needed to contain flood conditions
- the flow through the intake when the driving head is at flood level.

Calculation of wing wall height

In Figure 5.15, the flood level h_{flood} is related to the river flood flow Q_{flood} by the weir equation for a weir that spans the width of the stream (i.e. not a 'notched' weir as in section 4.5). The equation is:

$$Q_{flood} = 0.67\, C_w \sqrt{(2g)}\; L\; h_{flood}^{1.5} \quad (m^3/s)$$

where C_w is the coefficient of discharge and L is the width of the weir (in metres). By taking $C_w = 0.6$ and $g = 9.81m/s^2$,

$$Q_{flood} = 1.8\; L\; h_{flood}^{1.5}$$

Hence

$$h_{flood} = (Q_{flood} / 1.8L)^{0.67}$$

The value of Q_{flood} has to be obtained from records or estimated from known local conditions e.g. perhaps 10 to 20 times ADF. The height of the wing wall H_{wall} should be about 0.5m higher than the predicted flood level, so

$$H_{wall} = H_{weir} + h_{flood} + 0.5m$$

For instance, if the proposed weir is 10m wide and 1.5m high, and it has been estimated that Q_{flood} might at its worst reach $100m^3/s$, then

$$h_{flood} = (100 / (1.8 \times 10))^{0.67}$$

$$= \underline{3.1m}$$

so,

$$H_{wall} = 3.1 + 1.5 + 0.5$$

$$= \underline{5.1m}$$

If the weir had been 15m wide, the answers would have been:

$h_{flood} = 2.4m$ and $H_{wall} = 4.4m$

Thus a longer weir crest has the effect of lowering flood levels and reducing the height of the wing wall.

Calculation of flow through the intake

The flow rate into the channel is determined by the head of water above the centre-line of the intake, equal to $h_{flood} + h_i$, where h_i is the depth of the centre of the intake below the weir-crest (see Figure 5.15). The orifice equation then gives the water velocity through the intake as:

$$V_i = C_d \sqrt{(2g\,(h_{flood} + h_i))}$$

Continuing the previous example, if the intake is placed half way up between the river-bed and the weir-crest, then $h_i = 0.75m$. So for the weir of length 10m, taking $C_d = 0.6$,

$$V_i = 0.6 \sqrt{(2 \times 9.81 \times (3.1 + 0.75))}$$

$$= \underline{5.2m/s}$$

The volume flow through the intake is then simply $Q_i = A_i V_i$

The effect of changing the weir dimensions

If the weir-crest was lowered by 50% from 1.5m to 1.0m, making $h_i = 0.5m$, then the calculation gives $V_i = 5.0m/s$ (instead of 5.2m/s). In other words, changing h_i makes very little difference to the intake flow under severe flood conditions; under *normal* river conditions, however, lowering the crest would significantly *cut* the rate of flow through the intake. This is because the large figure of 3.1m for h_{flood} would be replaced by a height now less than h_i, say 0.4m. In which case V_i would drop from 2.9m/s to 2.5m/s.

On the other hand, widening the weir crest by 50% has a greater effect on the control of a flood. The height of the necessary wing wall is reduced from 5.1m to 4.4m, as we saw above, and V_i also decreases from 5.2m/s to 4.7m/s. Meanwhile the normal intake flow is barely affected, in fact the intake velocity would drop from 2.9m/s to 2.8m/s.

Siting the intake to avoid silting

The water of many rivers carries a significant load of silt, sometimes all the year round, but often only in certain seasons. There are two main problems caused by silt:

- if it is deposited it can eventually cause blockages of the flow and even diversion of the stream
- it can cause erosion and wear of penstocks, valves, sluice gates and turbine runners.

It is important to ensure that the intake is designed to avoid silt being deposited around it, which would eventually impede the flow to the turbine. Silt passing through the intake can be removed by incorporating settling basins into the headrace channel. Nevertheless, rivers carrying a heavy silt load may overburden the silt basins and clog the entrance to the penstock.

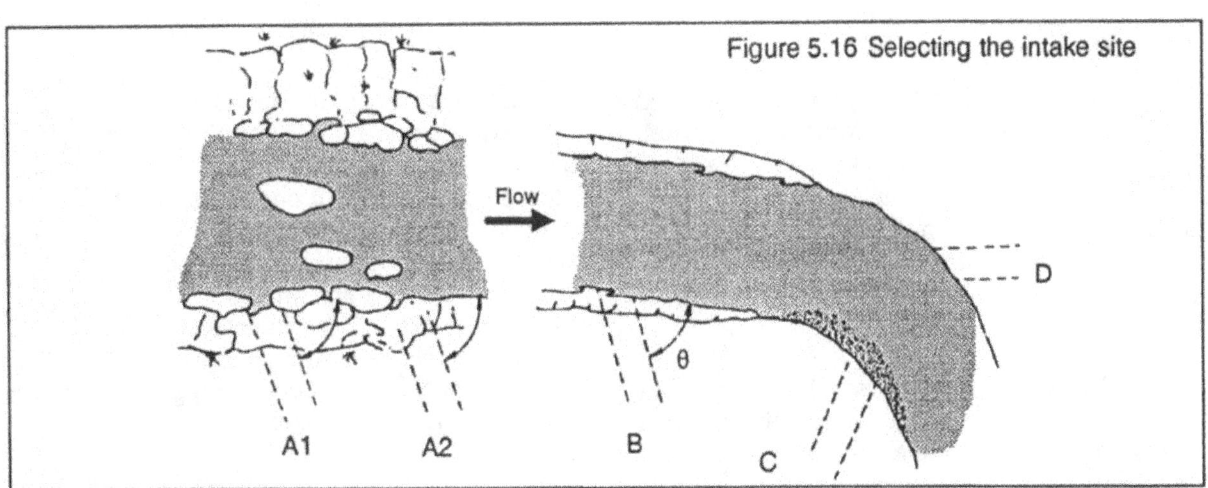

Figure 5.16 Selecting the intake site

Figure 5.16 shows a river in plan view. There is a mountainous and rocky section first, followed by a gentler section and a bend. Suppose that an intake site must be chosen somewhere along this stretch of river.

Sites A1 and A2 may pose difficulties with respect to access and civil works because of the steep embankment. Even when an intake is built, operation and maintenance could be problematic in high flow conditions. A1 is suitable for the placement of a weir anchored on boulders, assuming there is no danger of their moving in a flood. A2 offers the advantage of natural protection of the intake mouth. Silt is less likely to be a problem in rocky areas.

An intake sited in the bend at point C is unsuitable, since silt tends to deposit on the inside of a bend. This is due to a spiral flow effect under the surface of the water, as shown in Figure 5.17a. An intake in the straight section at B would also create a minor bend in the flow and set up the spiral movement which may cause silt to accumulate at the intake. One method of avoiding this effect is to deflect the flow on the other side of the river with a *deflecting groin*, as shown in Figure 5.17b. This can be added at a later stage once it is established whether or not silting of the intake is a problem. It may be sufficient to ensure that the angle θ is either very small, to avoid spiralling, or alternatively more than 80°, so that the side intake principle is used (see Figure 5.12). Site D is not recommended because it is liable to be damaged by boulders rolled downstream by flood flows, and it will collect floating debris very rapidly.

Skimmers

If floating debris is a problem, a steel or wooden bar (or 'skimmer'), can be positioned on the water surface at an angle to the flow so as to stop the debris and protect the intake.

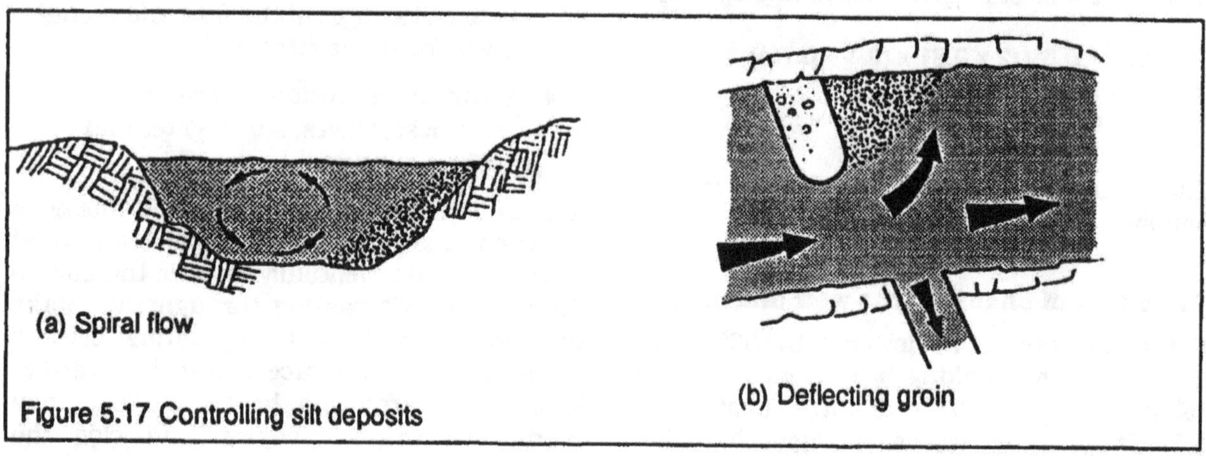

(a) Spiral flow

(b) Deflecting groin

Figure 5.17 Controlling silt deposits

5.6 Spillways

Spillways are designed to permit controlled overflow at certain points along the channel. Figure 5.18 depicts a flood spillway in detail, including flow control and channel-emptying gates. Flood flows through the intake can be twice the normal channel flow, so the spillway must be large enough to remove this excess flow.

The spillway is a flow regulator for the channel. In addition, it can be combined with control gates to provide a means of emptying the channel. In certain circumstances it is essential to stop the channel flow quickly, for instance if a break in the channel wall has occurred downstream causing progressive channel collapse and foundation erosion. Often emergencies will occur when the river is in flood. The control gates in Figure 5.18 must be placed above the flood line and at a distance from the stream as shown, to allow easy access when the river is in flood. It is sensible to rely not on one channel-emptying mechanism, but two, because a gate can jam, especially if not often used.

In general, gates are more easily opened than closed, so gates A and B in the figure could be more reliable than gate C. Nevertheless, if gate C is used to stop channel flow, the immediate effect will be for the full flood flow to pass over the spillway. The upstream channel width and wall height must be sufficient to accommodate the full back-flow in this case. Since C can act as a variable orifice, it is also a useful mechanism for regulating the flow to low levels if desired, e.g. for turbine testing.

The spill flow must be led back to the river in a controlled way so that it does not damage the foundations of the channel. Preferably a concrete and masonry drain should be built to provide a resilient passage for the spill flow. Rapid flows (> 4m/s) can be very erosive; to dissipate their energy rock protrusions or steps should be built into the bed of the spill drain.

Figure 5.18 Flood spillway

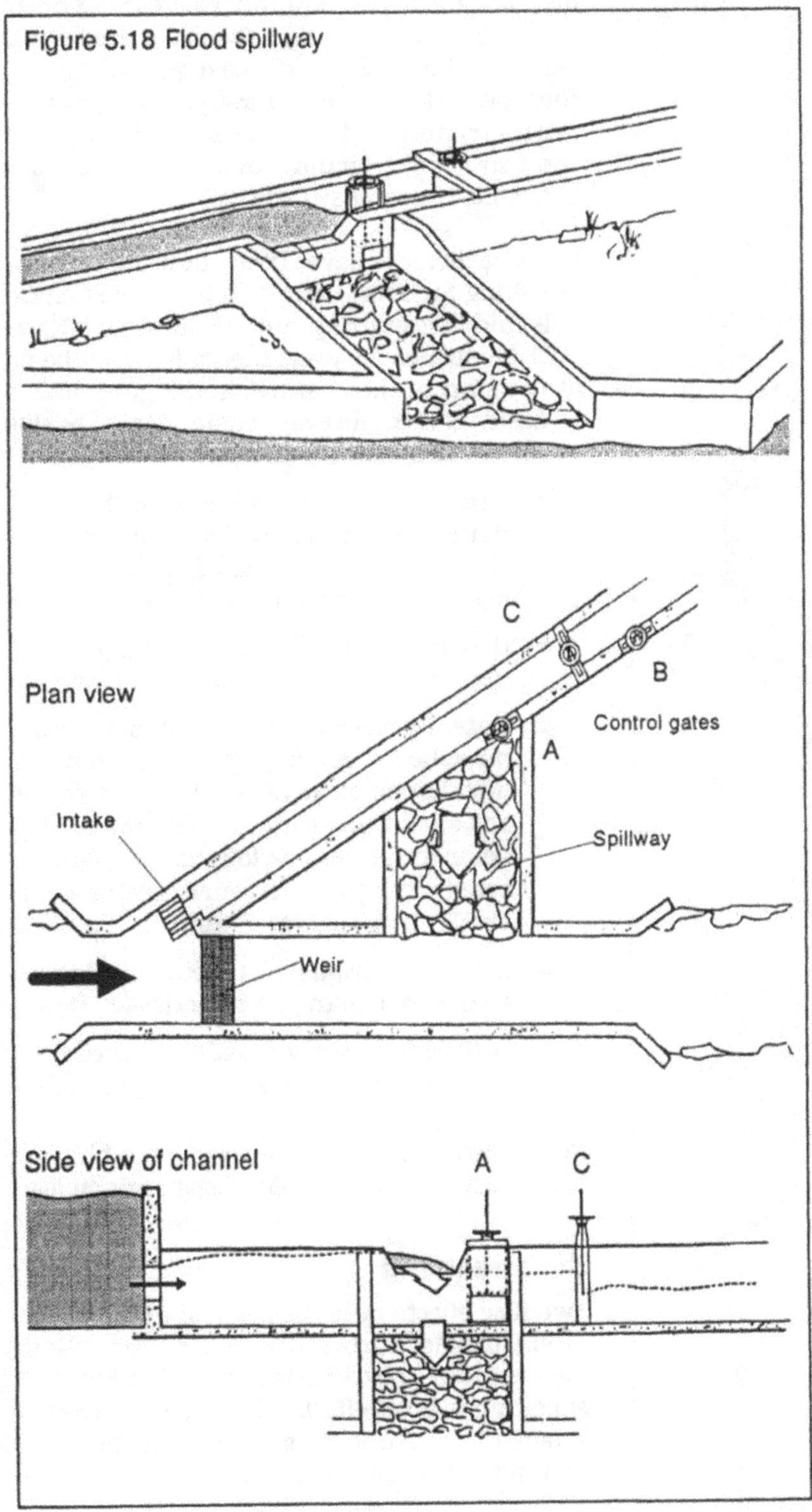

The height of the *spillcrest* should be aligned to the normal flow surface level so that overflow occurs at the first sign of flood (see Figure 5.19). The spillway must be wide enough to allow the complete spill flow to escape. The width required can be estimated by modelling the spillway as a weir and using the weir equation (see Section 4.4). As further width is usually cheap to provide, it is wise in practice to increase the width calculated.

It is possible to encourage excess flow to leave via the spillway by creating a back flow: for instance (see Figure 5.19) by installing a small weir crossing the channel bed, or by partially closing control gate C downstream of the spillway. Such obstructions should be profiled so that the water travels smoothly past them and does not become turbulent. No silt should be allowed to deposit in the spillway and control gate area. The flushing action induced by opening gate A can be an easy way of clearing silt deposits.

Further spillways are required at intervals throughout the length of the main channel as shown in Figure 5.20. These spillways prevent uncontrolled overflow of the channel walls in the case of a landslide or other blockage in the channel. They should be spaced typically every 100 metres, and have durable drains to lead the spilt water away from the channel foundations. The cost of including these spillways and drains is considerably less than the cost of repairing slope erosion and channel damage caused by an overflowing channel. However, on very stable, shallow slopes these spillways may be omitted.

Figure 5.19 Creating a back flow

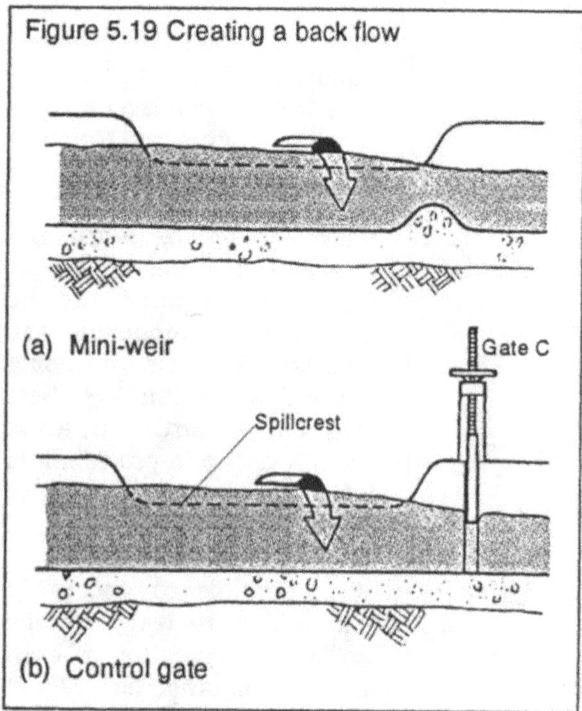

Figure 5.20 Spillway spacing; every 100m is advisable

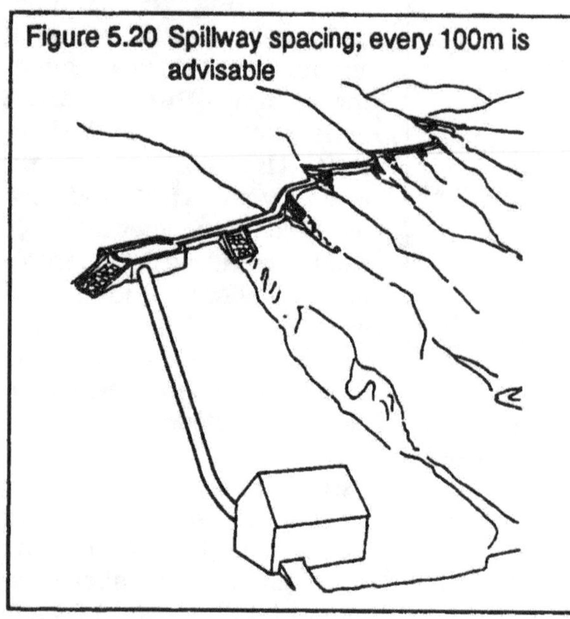

5.7 Settling Basins and Forebay Tanks

The water drawn from the river and fed to the turbine will usually carry a suspension of small particles. This sediment will be composed of hard abrasive materials such as sand which can cause expensive damage and rapid wear to turbine runners. To remove this material the water flow must be slowed down in settling basins so that the silt particles will settle on the basin floor. The deposit formed is then periodically flushed away. It is generally necessary to settle out the sediment both at the start of the channel and at the penstock entry, i.e. the forebay tank.

Many schemes have been built with poor de-silting facilities, resulting in clogged channels and fast-wearing turbine runners. Figure 5.21 illustrates a simple design for a silt basin at the start of the channel, and Figure 5.22 is a design for a forebay basin. Both basins should be designed as follows:

- they must have length and width dimensions which are large enough to slow the water sufficiently to cause settling of the sediments;
- they must allow for easy flushing out of deposits;
- water discharged from the flushing exit must be led carefully away from the installation so as to avoid erosion of the soil surrounding and supporting the basin and penstock foundations e.g. by constructing a paved surface with walls, similar to a spillway drain;
- they must avoid flow turbulence due to sharp area changes or bends;
- sufficient capacity must be allowed for collection of sediment.

Many variations in design are possible, but they must satisfy these five design principles.

Slipstreaming

Two flow effects must be avoided in the design of silt basins: turbulence and slipstreaming. Figure 5.23a shows an incorrect design which encourages both effects. Turbulence must be avoided because it stirs up the silt bed load and maintains silt in suspension. Slipstream-

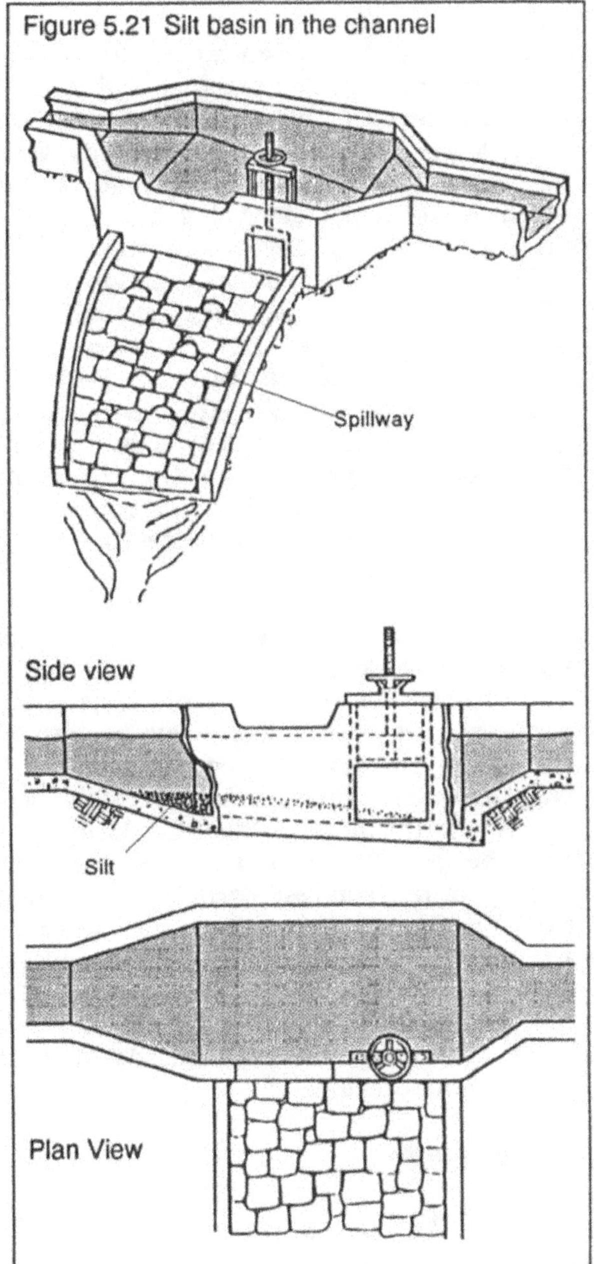

Figure 5.21 Silt basin in the channel

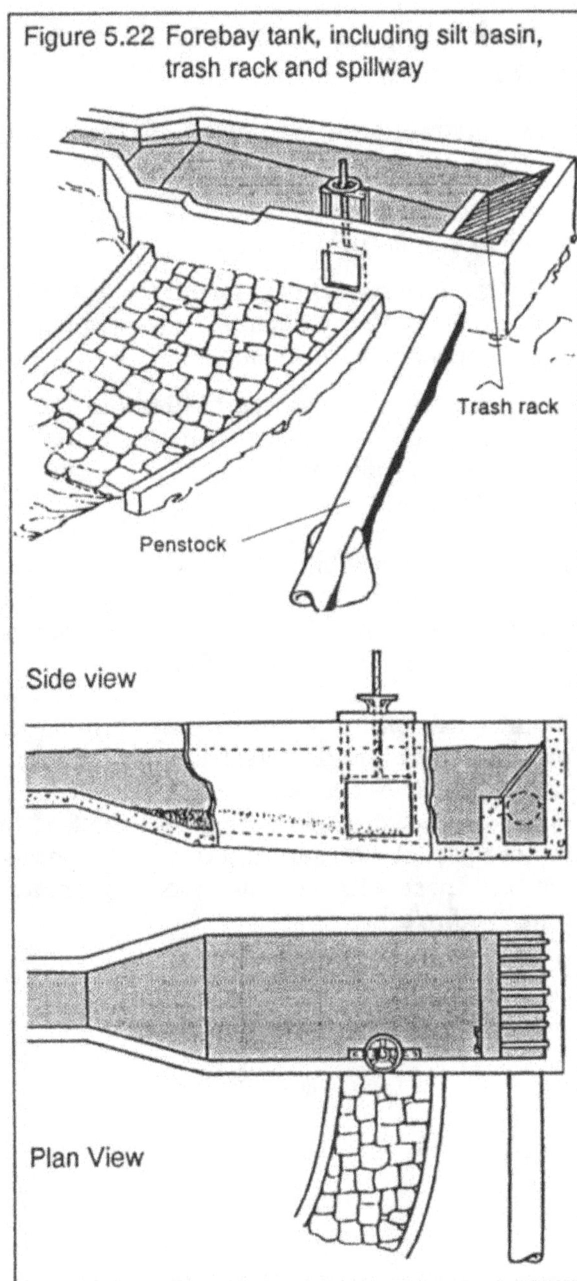

Figure 5.22 Forebay tank, including silt basin, trash rack and spillway

ing is the tendency for a body of water narrower than the total flow cross-section to move quickly through the basin from entry to exit, carrying a silt load with it. Figure 5.23b shows the entry and exit profiles needed to avoid this. As a further measure, the tank may have partitioning walls built into it in the direction of the streamlines to assist in smoothing the flow.

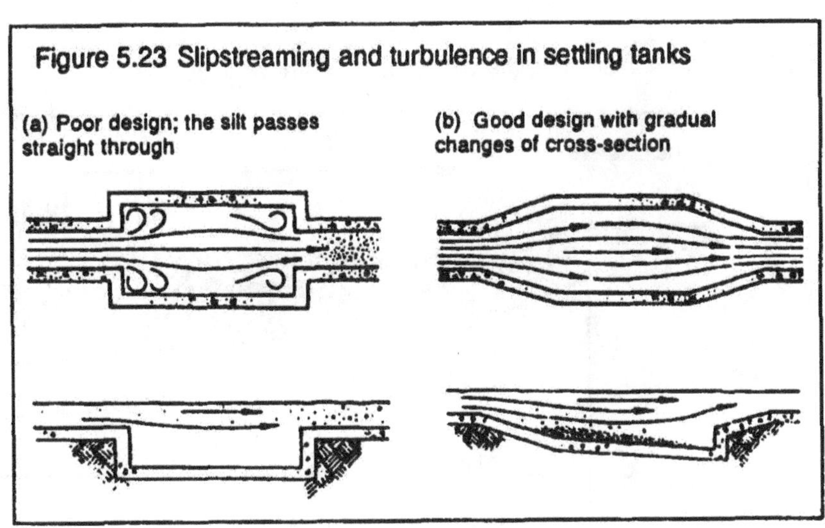

Figure 5.23 Slipstreaming and turbulence in settling tanks

Determining the dimensions of the settling area

The constraints on the design of the settling area are:

- the flow through it must be completely free from turbulence
- the flow must linger long enough in the tank for the sediment to have time to deposit out on the bottom
- the features of the particular location may set physical constraints on the length, width or depth of the settling tank.

Turbulence depends on the speed of the flow and the surface roughness. In general the speed of flow should not exceed 0.3-0.5m/s. The recommended value is 0.2m/s.

Smaller particles sink more slowly than larger particles. The settling velocities of different sizes of particles are tabulated in Table 5.1. In most micro-hydro schemes it is sufficient to remove particles bigger than 0.3mm in diameter, i.e. those with settling velocities greater than 0.03m/s.

The settling basin illustrated in Figure 5.24 is in three sections: the entrance, the settling area, and the exit. The entrance and exit are necessary to open the channel out smoothly to the larger dimensions of the settling region. The settling region itself is of length L and width W. Its depth is divided into two portions: D and D_{tank}. Silt accumulates in the collection tank at the bottom and must not be allowed to overflow beyond the height of D_{tank}, otherwise the basin will not operate correctly. The dimensions of the collection tank will depend on whether excavation is possible, where the silt-draining outlet will be, and how much silt is expected from the river. The method below describes how to calculate the range of settling tank dimensions that suit a specific situation.

Consider the worst possible case: a particle of silt has just arrived in the settling area and is furthest from the collection tank, right at the surface of the flow. If it is to be trapped, it must drop a vertical distance D while travelling a distance L through the tank. If the flow moves with the fastest allowable velocity V_{max}, and the particle drops vertically with the velocity of the smallest particle which must still be caught, u_{min}, then the particle will take a time L/V_{max} to reach the end of the settling area, and a time D/u_{min} to drop into the collecting tank. Therefore the condition for it to be trapped is that the second time is shorter than the first.

In other words:

$$\frac{D}{u_{min}} < \frac{L}{V_{max}}$$

Figure 5.24 Calculating the dimensions of the settling tank

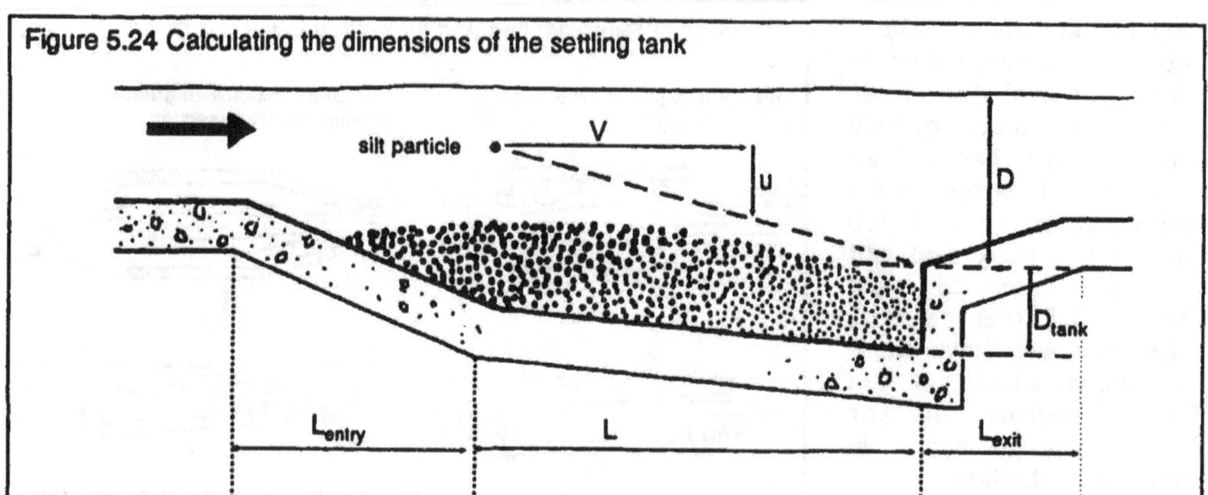

This can be rearranged as:

$$V_{max} D < u_{min} L \qquad \text{Inequality (i)}$$

Also, the flow velocity V is equal to the volume flow rate in the channel Q divided by the area cross-section of the settling tank (W x D). This velocity must be less than the velocity at which turbulence might occur, which is the maximum allowable velocity V_{max}. Therefore:

$$\frac{Q}{WD} < V_{max}$$

or

$$\frac{Q}{W} < V_{max} D \qquad \text{Inequality (ii)}$$

Inequalities (i) and (ii) can now be combined into a single expression:

$$\frac{Q}{W} < V_{max} D < u_{min} L$$

This simple inequality can be used to determine the possible settling tank sizes that would be appropriate for a given situation.

Example

As an example, suppose the rated flow is $0.25m^3/s$ (or 250*l*/s). The recommended settling tank velocity is 0.2m/s and the smallest particle that might damage the turbine is of size 0.3mm, so u_{min} is 0.03m/s.

If one of the settling tank dimensions (W,D or L) is now chosen, then the inequality above will determine the minimum sizes for the other two. For instance, if the basin depth D is chosen as 0.5m, then the inequality gives:

$$\frac{0.25}{W} < 0.2 \times 0.5 < 0.03 \times L$$

so $W > 2.5m$ and $L > 3.3m$

However it may be that the particular site is better suited to a basin of width 1.5m, and it is also thought safe to allow a faster flow without inducing turbulence. Then:

$$\frac{0.25}{1.5} < V_{max} \times 0.5 < 0.03 \times L$$

giving $L > 5.6m$ and $V_{max} = 0.33m/s$

The figures can be experimented with further until a compromise has been reached which is most applicable to the local conditions. As usual, it is advisable to increase the values of the calculated dimensions to allow for the simplifications involved in the mathematical model. A safety factor of 2 on the length of the basin is recommended.

Settling tank entry and exit profiles

To avoid turbulence in the water, exit and entry lengths (L_{entry} and L_{exit}) should be 2.5 times the settling area width W.

Collection tank capacity

The depth of the collection tank is specified on the basis of a reasonable emptying frequency. After periods of heavy rainfall when the silt load (or *turbidity*) is a maximum, a reasonable emptying frequency might be up to twice daily.

For instance, a channel flow of $0.2m^3/s$ carries a severe silt load of $0.5kg/m^3$. If 100% of the silt is deposited in the collection tank, then:

$$\text{Rate of silt deposited} = 0.5 \times 0.2 \text{ kg/s}$$

$$= \underline{0.1kg/s}$$

$$\text{silt deposited in 12 hrs} = 0.1 \times 12 \times 3600 \text{ kg}$$

$$= \underline{4320kg}$$

Taking the density of sand as $2600kg/m^3$, and estimating the packing density of the deposit as 50%, then:

$$\text{volume of silt deposited} = \frac{4320}{(0.5 \times 2600)} \; m^3$$

$$= \underline{3.3m^3}$$

If the collection tank has width 2.5m and length 6.6m, then its average depth needs to be:

$$D_{tank} = \frac{3.3}{(2.5 \times 6.6)} \; m$$

$$= \underline{0.2m}$$

In practice, the emptying frequency will depend on observations made of the collection tank-filling time. The forebay tank should, in theory, require a smaller collection capacity than the entry basin as by that stage the silt load of the channel should have been greatly reduced. However, channel water can easily become silted by the entry of debris into the channel, or as a result of failure of the entry basin. For this reason the forebay tank should be sized exactly as the entry basin and should be emptied as frequently as is observed necessary.

Emptying of silt basins

Usually emptying is a fairly tedious process. The sluice gate is opened and the sediment is shovelled along the basin floor to direct it through the gate. The basin floor is slightly sloped to facilitate this. During manual emptying, the intake control gate must be closed so that the water flow stops, which obviously requires shutting down the turbine.

A pipe-plug, or cylinder gate, as shown in Figure 5.25, allows sudden flushing out of the silt and is cheaper than a conventional sluice. It can be made from a short length of PVC or steel pipe.

Stop-logs or planks of wood located into grooves in the walls are not recommended as they leak in the dry season, are very difficult to remove for opening, and are easily stolen.

Assessing water turbidity

Turbidity is expressed as the weight of silt per m^3 of water. A simple method of measuring turbidity is to fill a bucket from the stream about 20 times at different depths. Wait each time until the silt has settled out, then separate and weigh the solid matter, also recording the volume of water collected. This is a very uncertain method unless there is some confidence that the worst part of the year has been chosen. Discussions with local people, and samples taken from irrigation channels, may be helpful. If other hydro schemes function nearby, it is essential that the operators are questioned about the frequency of desilting their settling basins. Design work can then be based upon these observations of existing basin design and performance.

Figure 5.25 Pipe plug for draining silt basins. Note that it also acts as a spillway

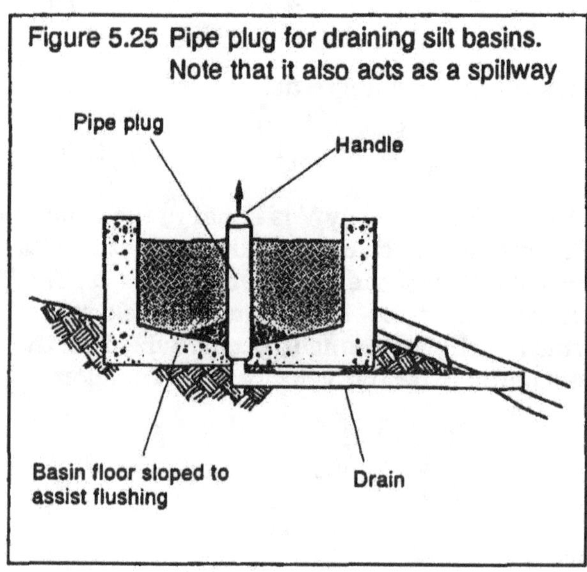

Screens

A screen (or trash rack) is a grill of vertically arranged parallel metal bars which intercepts floating debris as the flow passes through it. The channel flow is usually screened first at the intake and then at the entry to the penstock. The screen is a hindrance to the flow and introduces a slight head loss. Therefore the bar-spacing should be the maximum that will still trap debris large enough to damage the turbine.

In the case of a Pelton turbine, the bar spacing should not be more than half the nozzle diameter; but where a spear valve is used, it should be a quarter of the nozzle diameter. For a Francis turbine, the spacing should not exceed the distance between the runner blades. The head loss associated with a screen is best kept small by making the screen area large enough to reduce the speed through the screen to 0.2m/s. Also, the larger the screen, the less frequently it will block.

Cleaning screens

A screen can be cleaned either by removing it and knocking off the debris, or by using a specially made rake. The teeth of the rake should be spaced to fit between the bars, and the handle should be long enough to allow full raking of the screen. In some designs two screens are used in series so that they can be cleaned one at a time. Screens are best placed at an angle, roughly 60° from the horizontal (e.g. as in Figure 5.22), to assist the raking but also to allow gravity and buoyancy to keep the screen clear.

5.8 Channels

Figure 5.26 shows the various types of channel section which may be suitable for a particular installation. The various types considered are as follows, where 'sealing' refers to the application of a thin layer of material with no structural strength to reduce friction and leakage, and 'lining' refers to a material which adds structural strength to the channel walls:

- simple earth excavation, no seal or lining;
- earth excavation with seal (either cement slurry or clay);
- masonry lining or concrete channels;
- flumes or aquaducts made from: galvanized steel sheet, wood, pipes, pipes cut in half to form troughs, etc.

Channel parameters

The type of channel chosen for each part of the route is very important and some guidance on this is given below. Once the channel type and associated lining or sealing material have been selected for each section, it is possible to calculate suitable dimensions and the steepness of slope required. There are three governing parameters:

1. Water flow velocity
Excessively fast flow velocities will erode the channel, while a sluggish flow rate will result in silt deposition and clogging of the channel. Plant growth may also be a problem, and depends on local conditions. The worst case might require a minimum velocity of 0.7m/s. Table 5.2 gives recommended velocities for different channel materials. Within these constraints, higher velocities give cheaper channels but higher head losses, and hence reduce the power at the turbine.

2. Side slope
Water flowing in a channel made from loose material, such as sandy soil, will cause the walls to collapse inwards unless the sides are sloped and the width of the channel is large relative to its depth. Table 5.3 shows that stronger materials allow channels to have steeper walls.

The advantage of lining channels is that they can be narrower (and deeper) to carry the

Figure 5.26 Different types of channel

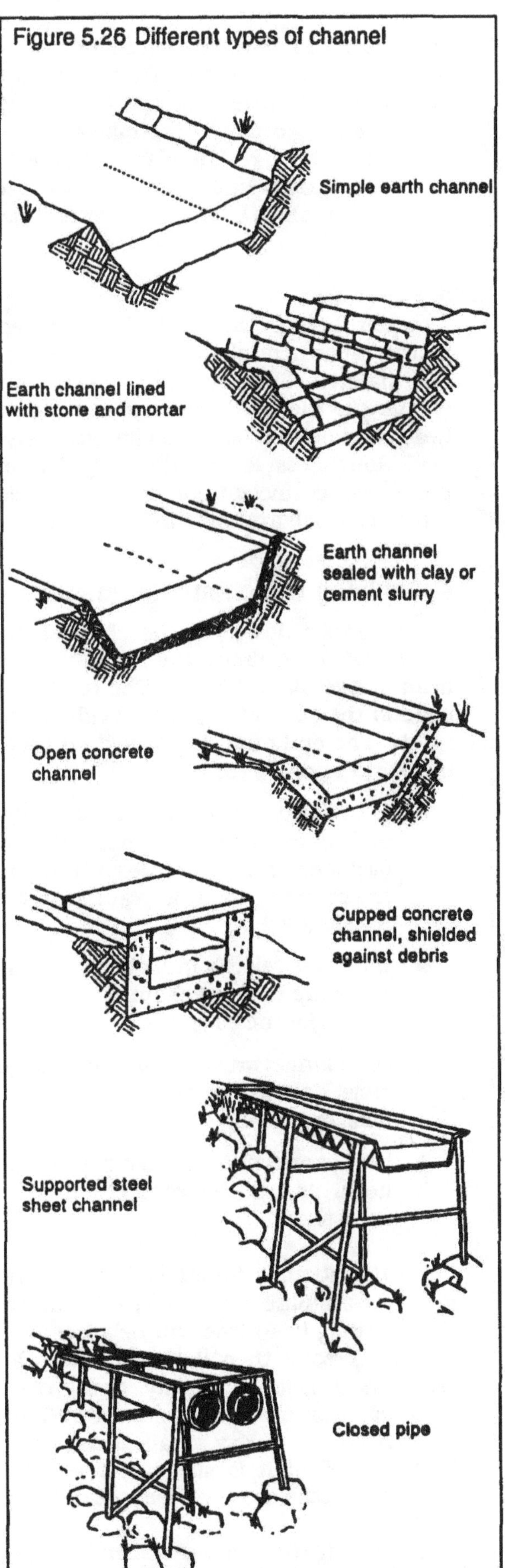

same flow, so less horizontal excavation is needed on difficult slopes. Trapezoidal profiles are normally chosen, except in the case of wood, steel, cement or masonry/cement channels, where rectangular ones are sometimes more easily built. Table 5.3 lists recommended side slope angles for different channel materials.

3. Roughness
As water flows in the channel, it loses energy in the process of sliding past the walls and bed material. The rougher the material, the greater the frictional loss, and the higher the head drop needed between channel entry and exit. Roughness is quantified by Manning's roughness coefficient for open channels 'n', values of which are listed in Table 5.4.

Optimizing the channel dimensions

A method for designing the channel dimensions and determining the slope and loss of head is given in Table 5.5. The total channel head is the sum of all the individual section heads. The optimum design will be governed by four key principles:

- the velocity of the water must be high enough to ensure that suspended solids (sediments) do not settle on the bed of the channel, and that plant growth is discouraged;
- the water velocity must be low enough to ensure that the channel walls are not eroded by the flow;
- the channel must be durable enough to resist destruction by storm run-off, rock falls, or landslip;
- the channel must be cost effective (i.e. no larger in cross-section than necessary).

The calculation is iterative. It starts with a guess of a suitable flow velocity from which the resulting head loss can be assessed. If it seems excessive then the calculation can be repeated for a lower velocity, giving rise to a larger channel cross-sectional area. Similarly, if the head loss is very small, the velocity can be increased (if it is still safe from causing erosion) and the channel size reduced.

In optimizing the channel design, it is important to compare the cost of a change in

Table 5.2 Maximum and minimum flow velocities

Material	Maximum velocity in m/s	
	Depth < 0.3m	Depth < 1.0m
Sand	0.3	0.5
Sandy loam	0.4	0.7
Loam	0.5	0.8
Clay loam	0.6	1.5
Clay	0.8	2.0
Masonry	1.5	2.0
Concrete	2.0	2.2

Water quality	Minimum velocity (m/s)
Clear	0.1
Silty	0.3

Table 5.3 Side slopes and hydraulic radii for trapezoidal channels

Material	Degrees from horizontal 'α'	Hydraulic radius 'r'
Sand	18	$0.271\sqrt{A}$
Sandy loam	27	$0.320\sqrt{A}$
Loam	34	$0.346\sqrt{A}$
Clay loam	45	$0.370\sqrt{A}$
Clay	60	$0.380\sqrt{A}$
Concrete	60	$0.380\sqrt{A}$

where 'A' is the cross-sectional area of the flow (or 'wetted' cross-section).

To find r for other values of α, the formula is:

$$r = 0.5\sqrt{\frac{\sin\alpha}{2 - \cos\alpha}} \times \sqrt{A}$$

In particular:
Rectangular cross-section: $\alpha = 90°$, $r = 0.35\sqrt{A}$
Triangular cross-section: $\alpha = 45°$, $r = 0.35\sqrt{A}$

Table 5.4 Manning's roughness coefficient 'n'

Type	n
Earth canals	
Clay	0.0130
Solid material, smooth	0.0167
Sand with some clay or broken rock	0.0200
Bottom of sand & gravel, with paved slopes	0.0213
Fine gravel, 10 - 30mm	0.0222
Medium gravel, 20 - 60mm	0.0250
Coarse gravel, 50 - 150mm	0.0286
Cloddy loam	0.0333
Lined with coarse stones	0.0370
Sand, loam or gravel, strongly overgrown	0.0455
Rock canals	
Medium coarse rock muck	0.0370
Rock muck from careful blasting	0.0455
Very coarse rock muck, great irregularities	0.0588
Masonry canals	
Brickwork, bricks, also clinker, well pointed	0.0125
Ashlars	0.0133
Thorough rubble masonry	0.0143
Normal masonry	0.0167
Normal (good) rubble masonry, hewn stones	0.0167
Coarse rubble masonry, stones roughly hewn	0.0200
Rubble walls, paved slopes, sand and gravel bed	0.0213
Concrete canals	
Smooth cement finish	0.0100
Fair-faced plaster	0.0109
Smoothed concrete	0.0111
Good formwork, smooth cement plaster, smooth concrete with high cement content	0.0118
Unplastered concrete, wood formwork used	0.0149
Tamped concrete with smooth surface	0.0161
Concrete shells with 150-200kg cement/m^3 according to age and design	0.0182
Coarse concrete lining	0.0182
Irregular concrete surfaces	0.0200
Wooden channels	
Planed, well-joined boards	0.0111
Unplaned boards	0.0125
Older wooden channels	0.0149
Natural water courses	
Natural river bed, solid and no irregularities	0.0244
Natural river bed, weedy	0.0313
Natural river bed with rubble and irregularities	0.0333
Torrent with coarse rubble (head-sized stones), bed load at rest	0.0385
Torrent with coarse rubble, bed load in motion	0.0500

channel speed with the cost of the corresponding change in the power output. For example, increasing the speed in a 100kW installation may save US$2000 in channel expenses, but it may also reduce the head available by 10%, or 10kW. $2000 is generally not a large sum to pay for an extra 10kW (providing it will be usefully consumed) so on balance it is probably better spent on a larger channel to allow a higher rated power output.

The method shown in Table 5.5 refers to the *hydraulic radius* (r) and the *freeboard allowance* (F). The hydraulic radius is a measure of the channel efficiency. An efficient channel is one in which as little of the water as possible is in contact with the channel surface, so inducing the least head loss due to surface roughness. The hydraulic radius (r) is equal to the cross-sectional area of the flow (A) divided by the *wetted perimeter* (P), where P is total wetted circumference of the channel e.g. in Figure 5.27, P = B + 2N. An efficient channel has a high hydraulic radius, and the most efficient profile is a semi-circle. The trapezoidal section is the most practical alternative to this and in Table 5.5 the dimensions of the trapezia that best approximate to semi-circles are calculated.

The freeboard is the amount the channel is oversized to allow for higher flows than the design flow. It is vital that the channel does not spill when it is carrying excessive water, because damage will rapidly occur to its walls and the hillside on which it is built. The usual freeboard allowance is 120%, meaning that the channel can accommodate 1.2 times the design flow.

Table 5.5 Method for calculating channel dimensions and head loss

Preliminaries:
Decide upon the length of channel L and the material with which it is made or lined. Also have a record of the rated channel flow Q.

Calculation steps:

1. Choose a suitable velocity V. Keep within the maximum and minimum velocities given in Table 5.2. Then calculate the wetted cross-sectional area $A = Q/V$.
2. From Table 5.3 find the side slope angle α and the hydraulic radius r.
3. Calculate the following:

   Wetted height $H = 2r$

   Wetted top width $T = 4r/\sin\alpha$

   Bottom width $B = T - 2H/\tan\alpha$
4. Find channel width and depth by multiplying T and H by the freeboard F (usually 1.2).
5. From Table 5.4 estimate n, the roughness coefficient of the wet surface. Calculate:

   $$\text{Slope } S = \frac{(nV)^2}{r^{1.33}} \quad \text{(Manning's Equation)}$$

   then Head Loss $= L \times S$
6. Repeat steps 1, 2, 3, 4, 5 for the other channel sections.
7. Add up the head losses for all the sections to find the total head loss. If too great or too small, repeat all steps with a different velocity.

Seepage loss

It is usually necessary to examine the soil along the route of the proposed channel. If the soil is very sandy it may be necessary to seal the channel against water loss by seepage. Either a sealant or a lining could be used for this purpose. A soil porosity test can be made using the technique summarized below, illustrated in Figure 5.28.

Push into the ground a cylinder of about 300mm diameter filled with water to a clearly marked level. Seepage into the soil reduces the level as time goes by. Each day or each hour, refill the cylinder and record the exact height added. Several of these cylinders can be used along the route. The records taken then provide an estimate of soil porosity. Table 5.6 lists typical seepage losses for different soil types. This test will provide a direct measure of the water lost should you decide to use this soil for a simple excavated channel without lining or sealing.

To calculate the seepage flow, take the factors from the table or from tests and use the wetted perimeter value P multiplied by the channel length L to calculate the area of wetted surface.

Unlined channel sections are considerably cheaper to construct. A survey of the route is necessary to detect where channel lining is likely to be needed, for instance in places where:

- the ground is excessively porous;
- rocky ground prevents excavation;
- the ground is steep and the soil unstable.

Fig 5.27 Channel dimensions and slope

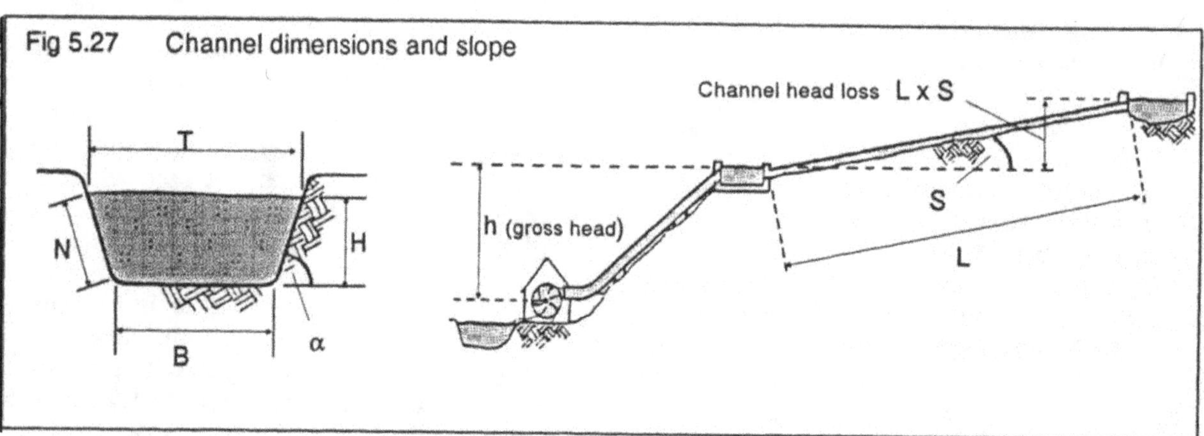

It is sometimes useful to inspect other channels in the area and consult local farmers with experience of irrigation channels. If the route passes through sandy soil where the seepage will be excessive, a length of pipe or a lined section of channel can be used. If the soil is waterlogged, the area can be drained using contour drains and lined main drains, or a pipe can be passed through the area. Rocks which obstruct the route can be removed, or blasted, or the route diverted around them.

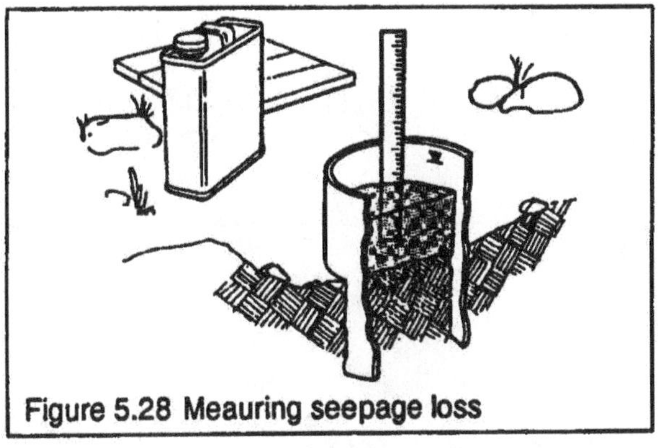

Figure 5.28 Meauring seepage loss

Lined channels allow water to be carried at a greater velocity because their walls are more resistant to erosion. This means that lined channels can be smaller in cross-section and a lot narrower and hence more economical in land use.

Table 5.6 Seepage loss

Soil type	Seepage losses (m^3/s per million m^2 wetted surface)	Equivalent Infiltration rate (mm/hour)
Sand	5.2 - 6.4	30+
Sand loam	3.5 - 5.2	20 - 30
Silt loam	2.5 - 3.5	10 - 20
Clay loam	1.5 - 2.5	5 - 10
Clay	0.5 - 1.5	1 - 5

A very important feature of an open channel is its vulnerability to damage from landslip and rock-falls, and from storm water run-off crossing its path. The cost of protection from these eventualities and their associated repair costs must be included in an estimation of the total channel cost. It may be that a low-pressure pipe to convey water, instead of an open channel, may be the cheaper option in the long-term because of savings in protection and maintenance. The use of a low-pressure pipe rather than an open channel will also save labour costs during construction; in particular where marshy ground and ravine crossings are a particular problem, the use of pipe can save the cost of aqueduct construction.

Channel crossings

Where small stormwater courses cross the path of the channel, great care must be taken to protect the channel. A heavy storm may create a torrent easily capable of washing the channel away. Provision of a culvert (a drain running the riverlet under the channel, as in Figure 5.29a) is usually not adequate protection. It will tend to block with mud or rocks just when most needed. In the long term it is economic to build a full crossing, as in Figure 5.29b. This should be sized to accommodate a flow of around 1000 times the usual wet season riverlet flow. These details are often neglected, resulting in channel damage, or at worst, large earthslips caused by the channel flow cutting into the hillside.

Figure 5.29 Channel crossings

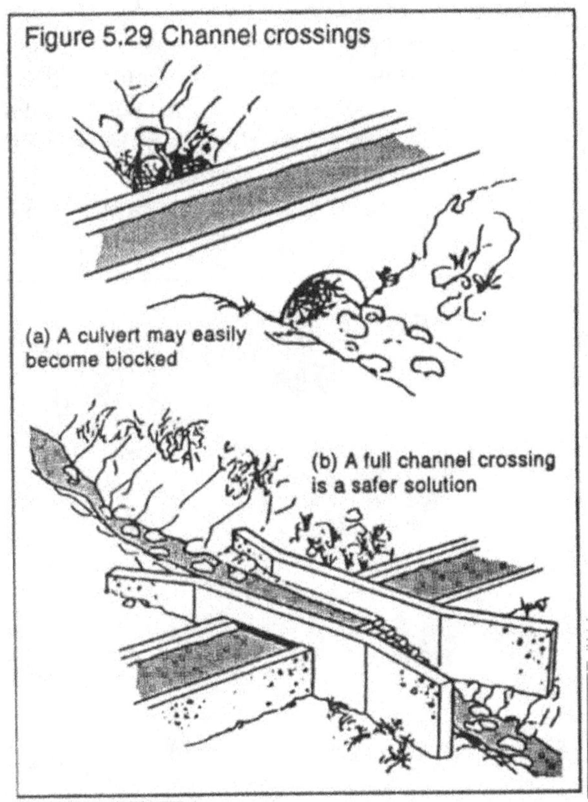

(a) A culvert may easily become blocked

(b) A full channel crossing is a safer solution

Penstocks 6

6.1 Introduction

THE penstock is the pipe which conveys water under pressure to the turbine. The major components of the penstock assembly are shown in Figure 6.1. The penstock often constitutes a major expense in the total micro-hydro budget, as much as 40% in high-head installations, and it is therefore worthwhile optimizing the design. The trade-off is between head loss and capital cost.

Head losses due to friction in the pipe decrease dramatically with increasing pipe diameter. Conversely, pipe costs increase steeply with diameter. Therefore a compromise between cost and performance is required, just as it was for channels.

The design philosophy is first to identify available pipe options, then to select a target head loss, 5% of the gross head being a good starting point. The details of pipes with losses close to this target are then tabulated and compared for cost effectiveness. A smaller penstock may save US$1000 in capital, but the extra head loss may account for lost revenue from generated electricity of US$500 per year.

It is never economic to lose *less* than 2% of the head in the penstock, so this sets one end of the design range. At the other end, at 33% head loss a given penstock is delivering the maximum hydraulic power possible for its dimensions. Any increase in flow would result in less power because the increased friction losses cause the effective head to decline more steeply than the increase in flow.

In projecting the cost of the penstock it is easy to underestimate the expense of peripheral items such as joints and even coats of paint. The difference in overall cost between one pipe material and another can be very significant if all these factors are included. Another factor often overlooked is transport implications. For example, a design may call for thin flanges which might easily be damaged during transport.

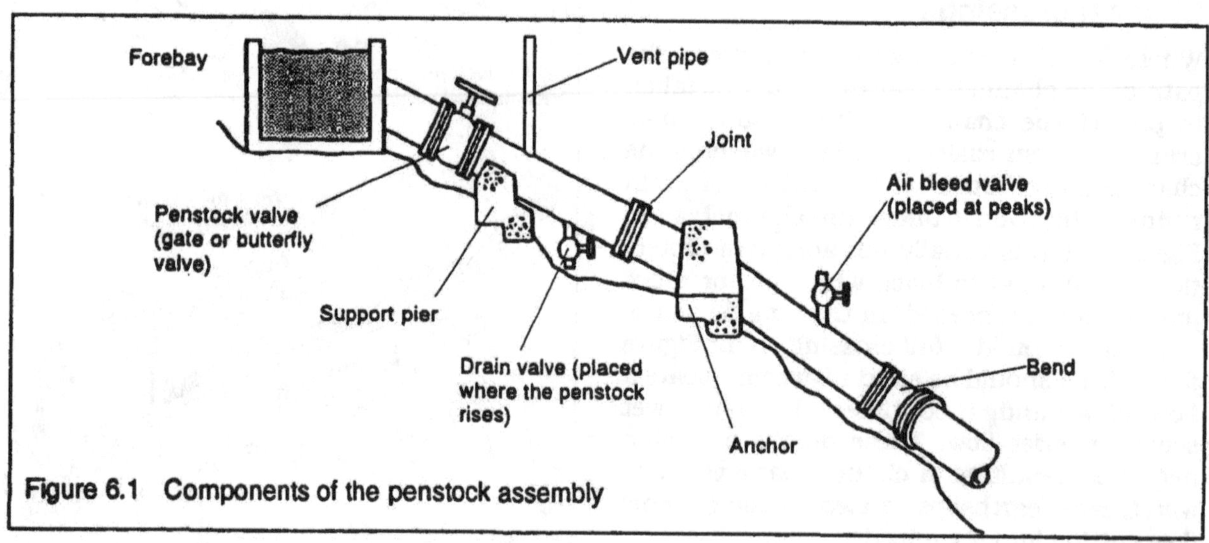

Figure 6.1 Components of the penstock assembly

6.2 Summary of Design Method

A methodical approach to the design of a penstock can be summarized in the following steps:

1. Consider the range of locally available materials and the possible types of joints. Compare maintenance implications and costs. Also list the diameters and wall thicknesses of pipes available on the local market.

2. Calculate the expected friction losses for a range of different pipe materials and internal diameters. Use 5% head loss as a starting point.

3. Add the extra 'turbulence' losses caused by bends, valves, etc. to the friction losses.

4. Predict the likely surge pressure in the case of accidental fast closure of the penstock valve, and add it to the static pressure. Calculate suitable wall thicknesses for the range of penstock sizes preferred.

5. Consider the number and type of supports, anchors, and joints for the types of penstocks preferred.

6. Estimate the overall cost of each option and check the availability of components from suppliers.

7. Compare the cost effectiveness of the most promising options.

8. Recalculate with 10%, 20% and 30% head loss to consider the use of smaller and cheaper pipe diameter to be used. Decide if the losses are acceptable and whether the savings in capital cost are sufficient to justify the losses. Note that to help save costs on a high-head system, lighter grades of pipe may be used at the top of the penstock where pressures are relatively low.

6.3 Materials

The following factors have to be considered when deciding which material to use for a particular project:

- surface roughness
- design pressure
- method of jointing
- weight and ease of installation
- accessibility of the site
- terrain
- soil type
- design life and maintenance
- weather conditions
- availability
- relative cost
- likelihood of structural damage

The following materials can be considered for use as penstock pipes in micro-hydro schemes:

- mild steel
- unplasticized polyvinyl chloride (uPVC)
- high density polyethylene (HDPE)
- spun ductile iron
- asbestos cement
- prestressed concrete
- wood stave
- glass reinforced plastic (GRP)

Mild steel, uPVC and HDPE are the most common materials used and will be discussed in greater detail. Table 6.1 summarizes the relative merits of each material.

Mild steel

Mild steel is perhaps the most widely used material for penstocks in micro-hydro schemes. It is relatively cheap, often available on the local market in a variety of sizes, and may be fabricated locally, requiring machinery common to most medium-sized steel fabrication workshops. It is made by rolling steel plate into a cylinder and welding the seam. It can be made in a variety of diameters and thicknesses as required. It has moderate friction loss characteristics, and provided it is well protected by paint or other surface coating can have a life of up to 20 years, depending on the mineral content of the water. Very soft water (low pH) can corrode steel pipes very quickly; buried penstocks are at the greatest risk. Mild steel pipes are resistant to mechanical damage and are relatively heavy, but they can be manufactured in lengths convenient for transport and installation if needed. They may be joined by bolted flanges, mechanical joints or by being welded on site. Cost savings can be made by selecting diameters which fit the standard sheet sizes available, thus saving on cutting and scrap.

uPVC

Unplasticized polyvinyl chloride (uPVC) pipe is becoming one of the most widely used alternatives to steel in micro-hydro schemes throughout the world. It is relatively cheap, widely available in a range of sizes from 25mm to over 500mm, and is suitable for high pressure use. Different pressure ratings are obtained by varying the wall thickness of the pipe, but generally the outside diameter remains constant for a range of pressure ratings in a given diameter, allowing several ratings in the same penstock. It is light, and easy to transport and lay. It has very low friction losses, does not corrode and is the easiest type of pipe to repair. It is relatively fragile, particularly at low temperatures, and prone to mechanical damage from falling rocks or from vehicles driving over it if it is only buried in a shallow trench. The main disadvantage is that uPVC deteriorates when subjected to ultraviolet light, causing surface cracking which can seriously affect the pressure rating. Therefore it must always be protected from direct sunlight, either by burying, covering with foliage, wrapping, or painting. This effect is more serious in hot climates and in the UK exposed penstocks show negligible deterioration after 10 years. Care is needed when burying uPVC pipes to ensure that sharp rocks or stones are not in contact with the pipe. Normally they are laid on a bed of sand and backfilled with sand or stone-free soil. uPVC pipes are jointed by mechanical spigot and socket, or by solvent welding. The latter is more difficult and requires special equipment to be done well, but is cheaper if many joints are involved.

Table 6.1 Comparison of common materials

★ = Poor, ★★★★★ = Excellent

Material	Friction	Weight	Corrosion	Cost	Jointing	Pressure
Ductile iron	★★★★	★	★★★★	★★	★★★★	★★★★
Asbestos cement	★★★	★★★★	★★★★	★★★	★★★	★
Concrete	★	★	★★★★★	★★★	★★★	★
Wood stave	★★★	★★★	★★★★	★★	★★★★	★★★
GRP	★★★★★	★★★★★	★★★	★	★★★★	★★★★★
uPVC	★★★★★	★★★★★	★★★★	★★★★	★★★★	★★★★★
Mild steel	★★★	★★★	★★★	★★★★	★★★★	★★★★★
HDPE	★★★★★	★★★★★	★★★★★	★★	★★	★★★★★

Prestressed concrete

Prestressed concrete pipes are generally suitable for use at up to 20m head. There are exceptions, however, and in China, prestressed concrete pipes are mass-produced for heads of up to 60m. These pipes are extremely heavy, and hence difficult to transport and lay.

Resistance to corrosion is good, but friction characteristics can vary from good to very poor. Jointing is generally by rubber ring joints. Quality control with regard to pressure rating can be a problem.

Spun ductile iron

Ductile iron has largely replaced cast iron, although the latter may still be found on older schemes. Ductile iron pipes are sometimes coated inside with cement, which gives them good protection from corrosion and a low friction loss. It is a heavy material, hence difficult to install, and tends to be costly.

Ductile iron pipes are usually joined by mechanical joints, push-in spigot and socket with a flexible seal, or occasionally flanged. These pipes are rarely cost effective, but are perhaps the strongest and most durable available, especially when concrete-lined, and can sometimes be found on the second-hand market.

Asbestos cement

Asbestos cement pipes are made from cement reinforced with asbestos fibres. They are unsuitable for use at heads greater than 20m, and are brittle, so transporting and laying them requires care. They are light, and have good friction loss characteristics. The dust caused by cutting asbestos pipes is a serious health hazard, so adequate protective clothing and equipment masks should be provided for people working with them. They are usually used with expensive steel and rubber couplings.

Wood stave

Wood stave pipe is made from strips of wood (generally pine or cedar) which are banded together by steel hoops, in much the same way as a barrel. Provided it is kept wet, woodstave has a long life, with a low friction level that may improve with age. Its very low weight means that it is easy to transport and makes the laying of the pipeline less complicated. However it is not suitable for high pressure applications. The main other disadvantage is its availability, since it is mainly manufactured in the USA, Canada and Scandinavia. It is enjoying a revival in popularity in recent years.

GRP

Glass-reinforced plastic (GRP) pipes are made from plastic resin, reinforced with glass fibre wound spirally and an inert filler. They are light, and very smooth inside, but are fragile, so care is needed during installation. They can be used at high pressures, provided they are buried in a trench which is backfilled with suitably fine material.

Joints are usually spigot and socket with a flexible seal. Depending on availability and relative cost, GRP may be a good option, but it has yet to gain wide acceptance. There is some evidence that GRP may be weakened by water absorption by osmosis over a long period.

HDPE

High density polyethylene (HDPE) pipes offer a good alternative to uPVC, though at a higher price. They are available in diameters from 25mm to over 1m. Small diameters (up to 100mm) are usually supplied in rolls of 50 or 100m, and are particularly useful for small schemes because of their ease of installation. HPDE has excellent friction loss and corrosion characteristics, and does not deteriorate as much as uPVC when subjected to sunlight.

HDPE pipes are generally jointed by heating the ends and fusing them under pressure. This requires special equipment, which is a disadvantage, but for smaller diameter pipes mechanical compression fitting joints are an economical option. HDPE has a coefficient of expansion ten times that of steel, so is often buried to keep the temperature steady.

6.4 Joints

Pipes are generally supplied in standard lengths, often 6 metres, and have to be joined together on site. There are many ways of doing this, and the following factors should be considered when choosing the best jointing system for a particular scheme:

- suitability for the particular pipe material
- skill level of personnel installing the pipes
- degree of joint flexibility required
- relative cost
- ease of installation (remembering weather conditions).

Methods of pipe jointing fall roughly into four categories, all of which are discussed below:

- flanged
- spigot and socket
- mechanical
- welded.

Flanged joints

Flanges are usually only used with mild steel pipes. They are fitted to both ends of a pipe during manufacture, and each flange is bolted to the next during installation. A gasket or other packing material, usually rubber, is necessary between each pair of flanges (see Figure 6.2). Flange-jointed pipes are easy to install, but the flanges add to the cost of the pipe. Mild steel pipes are generally flanged, but ductile iron ones only occasionally. The flanges should conform to some recognized standard e.g. British Standard (BS) or International Standards Organisation (ISO).

Flange joints are inflexible because they are designed to be strong enough take bending stresses. On low-head schemes this can cause flange costs to be equal to pipe costs. However, many current steel penstocks employ light-weight flanges, typically twice pipe thickness, and rely on good pipe support to reduce the bending stresses. This approach has gained wide acceptance in Nepal and Sri Lanka, for example, with significant cost savings, but there is no official standard for it.

Spigot and socket

Spigot and socket joints (Figure 6.3) require one end of each pipe to form a socket for the next pipe to fit into. Either the diameter of the end of the pipe is increased during manufacture, or the end is fitted with a collar. The plain end of each pipe can thus be pushed into the collar or 'socket' of the next. A good seal is required between each pipe section; depending on the pipe material this is either (a) a rubber seal, or (b) solvent cement adhesive.

Rubber seal

Rubber seal joints are generally of two types: O-ring seals or single/multiple 'Vee' lip seals. They are generally used to join ductile iron, PVC or GRP pipes. Both types permit a few degrees of deflection.

A few precautions are necessary when installing this type of joint:

- the seal must be clean before assembly;
- a special lubricant must be used because an oil-based grease will rot the seal; washing-up liquid is a possible solution;
- pipe clamps and a ratchet puller may be necessary to pull the joint together; essential for diameters over 150mm;
- ensure that the joint is properly aligned before final coupling; 'Vee' lip joints are extremely difficult to take apart;
- the pipes should not be too cold when joined or the pipeline will buckle when it warms up. If too hot, the joints may open up when the material cools and contracts.

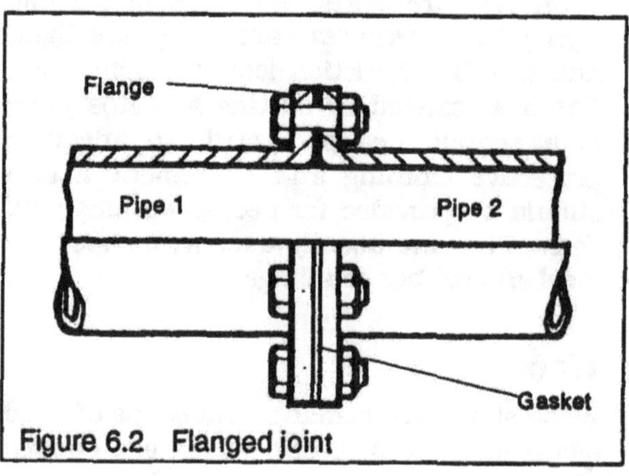

Figure 6.2 Flanged joint

Solvent cement seal

Solvent cement welded spigot and socket joints are used on uPVC pipes. The solvent, which dissolves the plastic material, is applied to the spigot half of the joint which is then inserted into the socket so that the two halves weld together. The following precautions should be observed:

- the joint must be scrupulously clean and dry, so do not attempt this method of jointing when it is raining; use special solvent cleaning fluid and a clean rag, and after cleaning, the joint should not be touched as even finger grease will prevent a proper bond;
- clamps and a ratchet puller should be used to control the jointing on pipe diameters greater than 150mm;
- align the joint with great care because it is IMPOSSIBLE to take solvent welded joints apart, even after a few seconds;
- the fumes from the cement are highly toxic and flammable; avoid prolonged exposure to them, and ensure adequate ventilation;

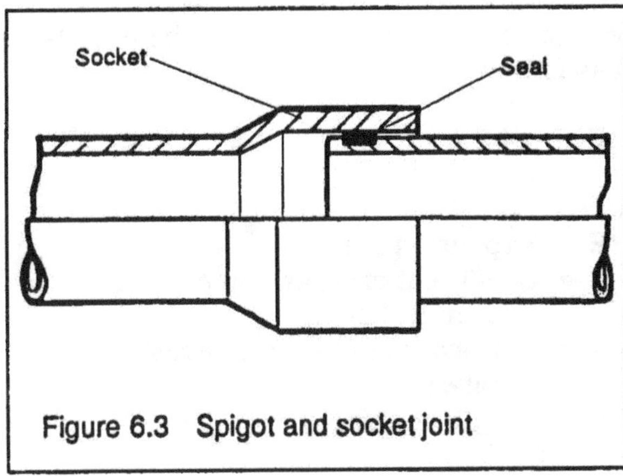

Figure 6.3 Spigot and socket joint

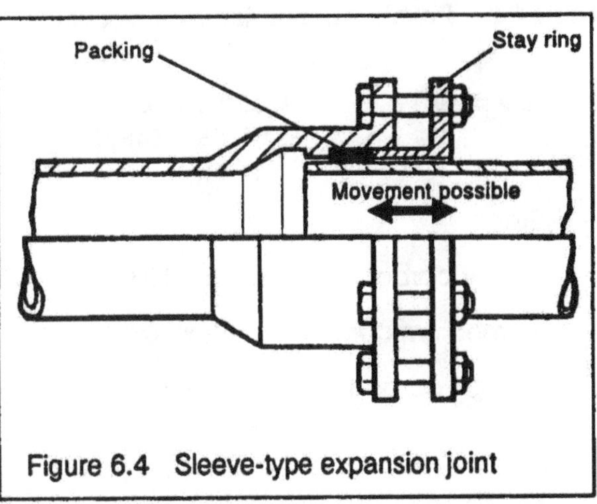

Figure 6.4 Sleeve-type expansion joint

Mechanical joints

Mechanical joints such as Viking Johnson (VJ) couplings are rarely used on penstocks because of their cost. Their main application is for joining pipes of different materials (e.g. mild steel to uPVC) or where a slight deflection in the penstock is required that does not warrant installing a bend. They may also be used in repair work to attach a replacement section of PVC pipe. Some types of mechanical joint cannot take any strain in the direction of the pipe, and have to be restrained by anchor blocks.

Welded joints

Joints can be welded on mild steel penstocks, and special techniques can also be used to weld HDPE and uPVC. Mild steel pipes are welded together on site, generally using an arc welder. It is a relatively cheap method, but has the disadvantage of needing skilled site personnel and an arc welder and generator set in remote and difficult terrain. It is essential to have a competent person doing the welding. Deflections can easily be accommodated using welded joints.

Fusion welding

HDPE or uPVC pipes can be joined by *fusion welding*. This also requires specialist skills and expensive equipment (the equipment can often be rented from the pipe manufacturer) but is a very rapid process. The two halves of the joint are fixed in a special jig, and heating pads applied to each end. The temperature of the heating pad and the length of time it is applied are critical in obtaining a good joint. When the material at the end of each pipe is semi-liquid, the two ends are brought together with some force, causing them to fuse together.

Expansion joints

Expansion joints such as the sleeve-type joint in Figure 6.4 must be included in steel and PVC penstocks unless the pipeline is buried. One expansion joint is required between each anchor block.

The expansion of the pipeline can be calculated as follows:

$$E = \alpha\ \delta T\ L$$

where:

E = Expansion (m)
α = Coefficient of expansion of steel (see Table 6.2)
δT = Temperature change experienced by pipe (°C)
L = Length of pipe (m)

For instance, if a steel penstock 87m long experiences a 20°C temperature change between night and day, then:

$$E = 0.000012 \times 20 \times 87$$

$$= \underline{0.021m} \quad \text{(or 21mm)}$$

To be safe, it would be recommended that the expansion joint be capable of accommodating a length change of double this amount i.e. about 40mm.

Table 6.2 Physical characteristics of common materials

Material	Coefficient of linear expansion (m/m°C)	Density (kg/m³)	Ultimate tensile strength (N/m²)	Young's modulus (N/m²)
Steel:				
low carbon	12×10^{-6}	8000	430×10^{6}	207×10^{9}
medium carbon	12×10^{-6}	8000	480×10^{6}	207×10^{9}
high carbon	12×10^{-6}	8000	$620 - 850 \times 10^{6}$	207×10^{9}
Ductile iron	11×10^{-6}	7100	$310 - 520 \times 10^{6}$	170×10^{9}
uPVC	54×10^{-6}	1400	27.5×10^{6}	2.75×10^{9}
HDPE	140×10^{-6}	960	$5.9 - 8.8 \times 10^{6}$	$0.19 - 0.78 \times 10^{9}$

6.5 Valves

A valve is often sited at the base of the penstock where it is very convenient for cutting off the flow of water to the turbine. However a valve should also be sited at the inlet to the penstock because there is a risk of damaging the pipeline with a pressure surge if the valve at the base is closed too quickly.

There are numerous types in use. For micro-hydro applications we will limit our discussions to gate and butterfly valves.

Gate valves

A gate valve (Figure 6.5) is a disc, larger in area than the flow, which can slide up and down within its housing to obstruct the flow as required. A closed gate valve at the base of the penstock can be very difficult to open again because of the huge pressure difference across it so they are best not used where large pressure differences occur.

Butterfly valves

A butterfly valve (Figure 6.6) is a lens-shaped disc which rotates on a central shaft. Little force is required to operate it because the upstream pressure is equally split either side of the pivot and so is approximately balanced during opening and closing. This means that the valve can be closed very quickly. However,

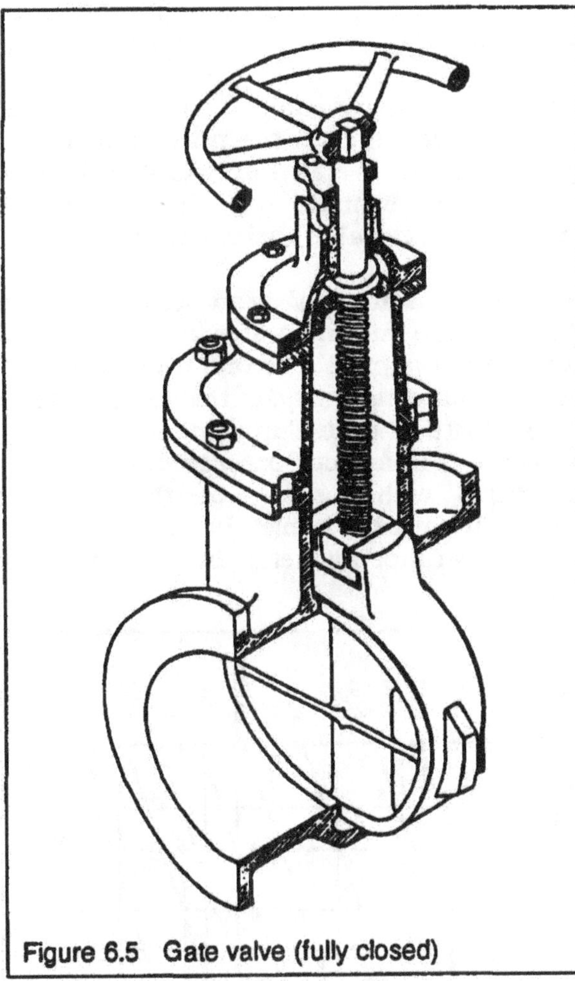

Figure 6.5 Gate valve (fully closed)

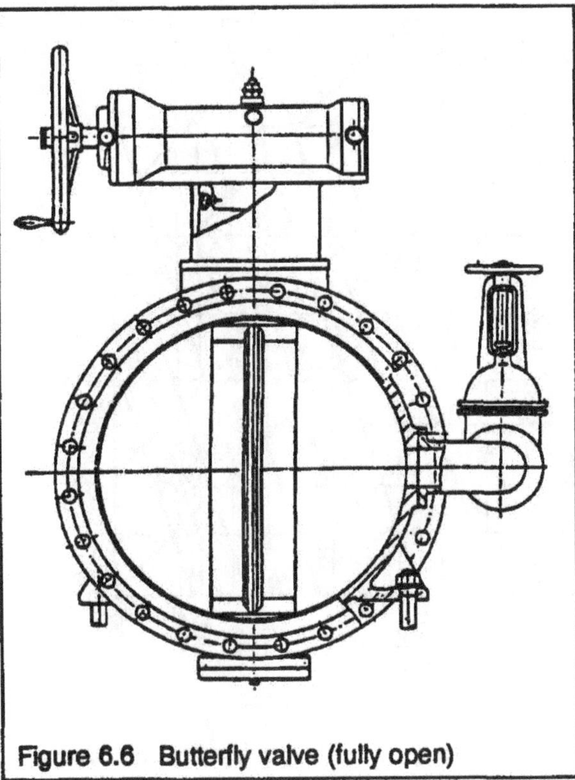

Figure 6.6 Butterfly valve (fully open)

decelerating the flow too rapidly will create a huge pressure surge in the penstock (the 'water hammer' effect). The penstock can actually burst or move its anchorage in reaction to the large forces that arise. For this reason it is not good practice to use butterfly valves unless they have low-geared closing mechanisms to prevent rapid closure.

Low-cost alternatives

In many small schemes the decision is made not to include a high pressure cut-off valve at the powerhouse in order to save on capital expenditure and to avoid the risk of damaging the penstock by closing the valve too quickly. This has an obvious safety penalty and cannot be recommended in general. If, for example, the turbine control valves become clogged or obstructed there is no quick way of cutting off the flow and the penstock may be ruptured by the surge in pressure.

Valves fitted at the forebay can be cheaper as they work on low pressure. They are anyway essential for draining the penstock. Simple sluices or pipe plugs can be used (see Section 5.7). The lowest cost alternative is a rubber sheet (e.g. an inner tube) stapled to a suitable piece of wood, and positioned over the protruding end of the penstock pipe. The rubber deforms under pressure producing a good seal. A vent pipe must be fitted at the top of the penstock to allow air in so that the water in the penstock can drain down.

Air bleed valves

Any penstock which has variations in slope needs air bleed valves at each high spot to allow trapped air to be removed because air in the penstock can severely restrict the flow. The simplest method of bleeding the air is to drill a hole in the pipe, usually at a joint, and use a self-tapping screw to close the hole once the air has escaped. A better arrangement is to purchase automatic air bleed valves.

Drain valves

A penstock which follows a rising and falling path will also require a drain valve at each low point in the path so that if necessary the penstock can be completely emptied of water. Without a drain valve, water would collect in the troughs in the pipeline.

6.6 Sizing and Costing

The following section describes how to select a penstock pipe with a suitable diameter and wall thickness. This optimizing process can be completed rapidly while considering the feasibility of a scheme. It can also be used for the final selection of the penstock, in which case it is a good idea to draw out the penstock route in both profile and plan on the basis of a careful site survey. Notes should be taken of soil and rock structure, obstructions, changes of horizontal direction, and gradient. These will allow bends, supports and anchors to be included in the costing process and will provide information needed for the friction loss calculation.

Pipe diameter

The two main constraints on the choice of pipe diameter are (a) the price, and (b) the head loss. The overall aim is to find the pipe of smallest diameter (i.e. cheapest) which causes an acceptable level of head loss.

The major source of head loss is friction between the moving water and the inside surface of the pipe. A second cause is turbulence. Both cause some of the water's energy to dissipate as heat. Turbulence losses occur whenever the smooth passage of the flow is interfered with, usually at the following places: the penstock inlet, bends, valves, and changes in pipe diameter. In very long pen-

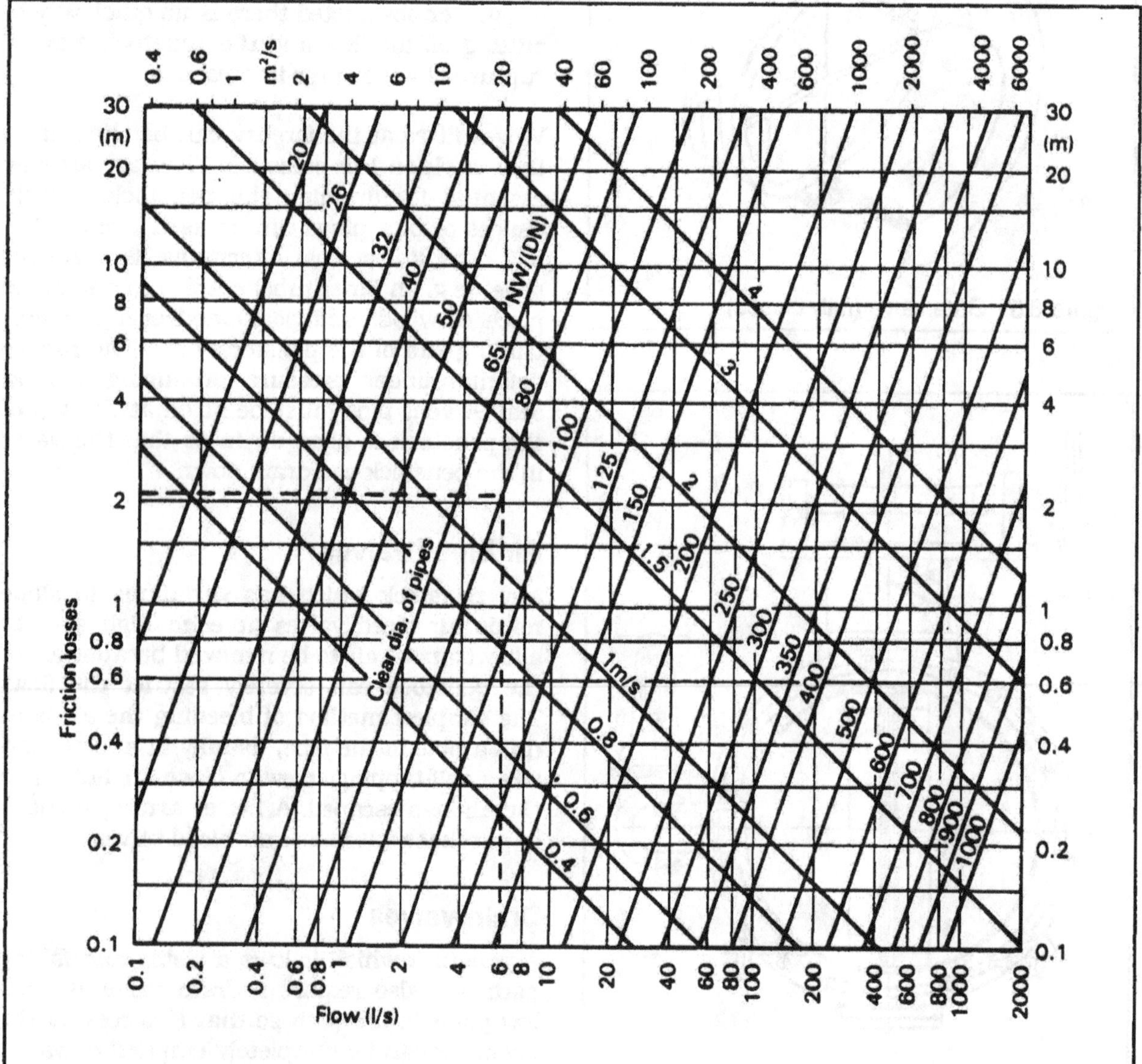

Figure 6.7 Friction losses in metres of head per 100m of new cast iron pipeline. Multiply by 0.8 for new rolled steel or plastic, or 1.25 for rusty cast iron.

stocks these losses are insignificant, but in short penstocks they may be greater than the friction loss.

In choosing a pipe diameter the objective is to restrict the total head loss to between 2% and 10% of the gross head. A reasonable target to aim for is 5%. For very long, high-head penstocks up to 33% head loss may be considered, especially in conjunction with multi-jet Pelton wheels.

Selecting the diameter is an iterative process. It involves starting with a first estimate of what might be a suitable diameter, given the sizes that are available, and then adjusting that estimate in the light of the calculated head losses and the price. The procedures for calculating (a) the friction loss and (b) the turbulence loss are described below. Note that if the calculation is only meant to be a rough estimate, then ignore the turbulence losses unless the penstock is very short.

Calculation of friction loss

Once the flow rate Q and the length of penstock L have been determined (or estimated), the friction loss depends only upon the diameter of the pipe D and the roughness of the pipe surface.

Friction loss in pipes is a complex topic. Some texts have recommended using Manning's Equation (as used in Table 5.5) but in fact this formula is only applicable to open channels and *not* closed pipes. Fortunately a friction loss calculation can often be avoided because pipe manufacturers tend to publish graphs or tables of friction loss at different flows. For instance Figure 6.7 is a friction loss chart for cast iron pipes. It can be applied to other materials by using the conversion factors listed at the bottom. The dotted line on the chart shows that an 80mm diameter pipe carrying a flow of 6*l*/s will cause a head loss of about 2m in every 100m.

If such a chart is not available, the recommended alternative is to use an equation formulated by Darcy, together with a chart prepared by Moody (Figure 6.8). Darcy's equation is as follows:

$$\text{Friction head loss} \quad h_f = \frac{2 V^2 L f}{g D}$$

where:

V = flow velocity (m/s)
L = penstock length (m)
D = pipe internal diameter (m)
f = friction factor

The value of the friction factor f depends mainly upon the surface roughness of the pipe material, but it also varies slightly with the Reynolds Number of the flow. The Reynolds Number Re is a product of different variables which has been found to be useful for describing a flow with a single number. For a flow of cold water in a pipe, the Reynolds number is simply

$$Re = V D \times 10^6$$

The magnitude of f in a particular situation is found using Moody's chart and an estimate of k, the ***equivalent grain size*** of the material. k has evolved from the initial experiments in which grains of sand of known sizes were stuck onto pipe surfaces. The roughness of any surface is now related to those artificially produced roughnesses by k, the 'equivalent' size of sand grains that would have the same effect. Values of k (in mm) for different materials are listed in Table 6.3.

To find f from the Moody chart, first calculate the Reynolds number of the flow and the ratio k/D (*Note:* convert k into metres first). Then, on the right-hand side of the Moody chart, find the curve which starts at your value of

Table 6.3 Equivalent grain size 'k'

Material	k (mm)
Smooth welded steel	0.015 - 0.3
Riveted steel	1.0
Rusted steel	2.0
Cast iron	0.15 - 30
Very smooth concrete	0.06 - 0.6
Unfinished concrete	3.0
Planed wood	0.3
Rough wood	0.7
Plastic tubing	0.003

k/D. Follow it to the left until you reach the vertical line corresponding to the Reynolds number of the flow. Then read off the value of f from the left-hand side of the chart. Use this value of f in Darcy's Equation to find the friction head loss h_f.

Example

A new PVC penstock, 500m long and with an internal diameter of 246mm, has a design flow of 150*l*/s. What is the head loss?

$$\text{Flow velocity } V = \frac{4Q}{\pi D^2}$$

$$= \underline{3.15\text{m/s}}$$

$$\text{Reynolds number Re} = V D \times 10^6$$

$$= \underline{7.7 \times 10^5}$$

k for PVC is given as 0.003, so:

$$\frac{k}{D} = \frac{0.003}{246}$$

$$= \underline{1.2 \times 10^{-5}}$$

On the Moody chart in Figure 6.8, following the curve of k/d = 0.00001 until it reaches Re = 8 x 10⁶ leads to a friction factor f = 0.0023.

Then using Darcy's equation,

$$h_f = \frac{2 \times 3.15 \times 500 \times 0.0023}{9.81 \times 0.246}$$

$$= \underline{9.6\text{m}}$$

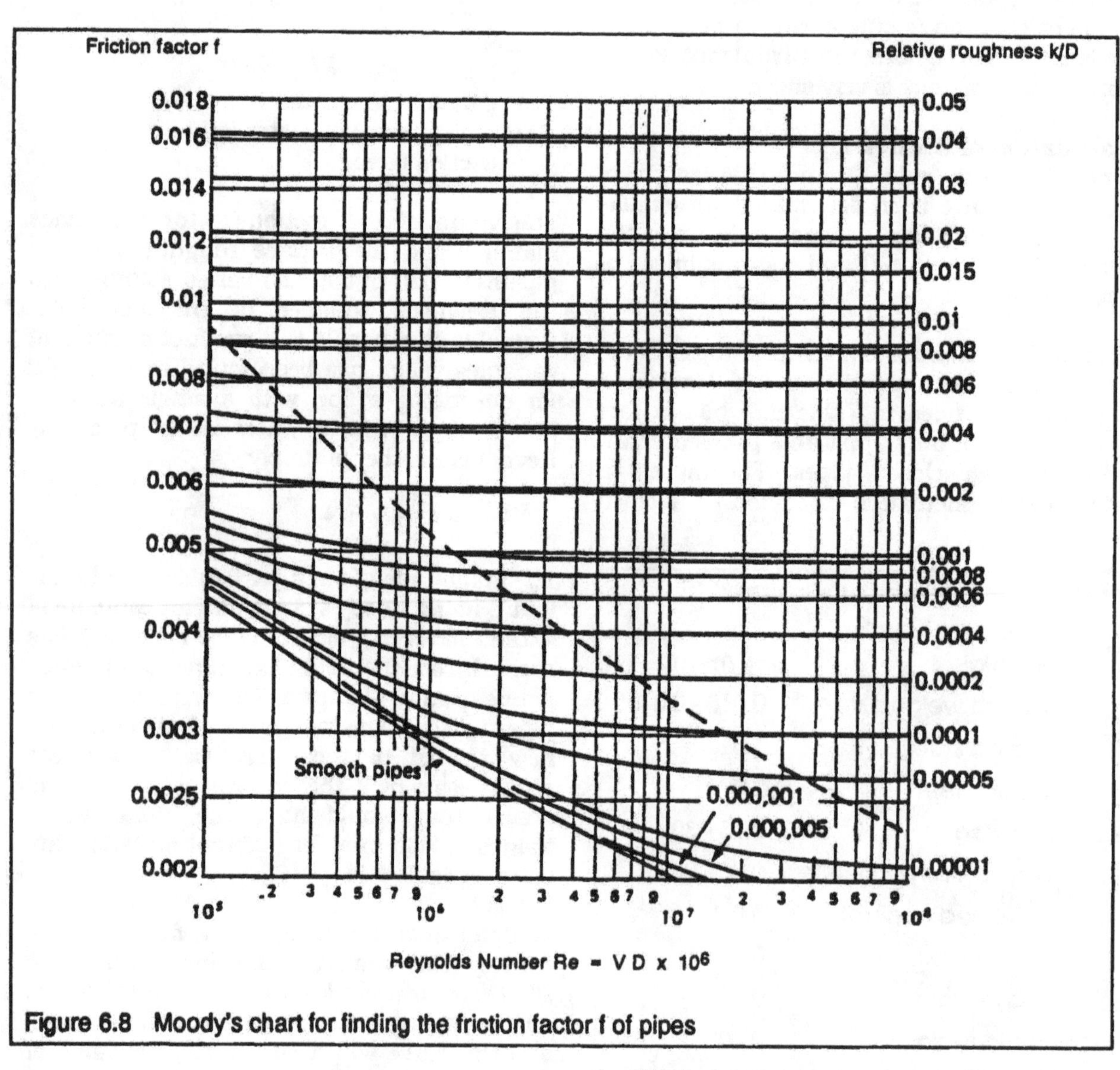

Figure 6.8 Moody's chart for finding the friction factor f of pipes

Calculation of turbulence losses

The magnitude of the turbulence losses depends upon the various changes which occur in the penstock geometry and the speed with which the flow hits those changes. The overall head loss due to turbulence is generally expressed as:

$$h_t = \Sigma K_i \frac{V_i^2}{2g}$$

where V_i is the velocity of the flow *after* it has encountered a feature with turbulence loss coefficient K_i. (Σ denotes "sum of".)

Table 6.4 shows how to determine the turbulence loss coefficients for various features, including intake profiles, angles of bend, pipe contractions, and types of valve. These coefficients are very approximate and are only useful for obtaining an order-of-magnitude estimate of the head loss. The use of turbulence loss coefficients is illustrated with the example in the box opposite.

The flowchart in Figure 6.9 summarizes the procedure for calculating friction and turbulence losses, and hence for determining the correct pipe diameter for a chosen material.

Pipe thickness

The penstock pipe wall has to be thick enough to withstand the maximum water pressure that might occur. This means not only coping with the normal operating pressures but also with surge pressures. Surge pressures are caused by sudden changes in flow velocity. They are generally short lived, but they can be very large. They occur whenever the valve at the base of the penstock is opened or closed, or if there is an accidental blockage such as a piece of debris lodged in a Pelton wheel nozzle.

The surge pressure adds to the static pressure, which varies linearly with height from zero at the penstock inlet to the gross head (h_{gross}) at the base. The magnitude of the surge depends upon the time taken to slow down the flow. Here we will only consider the worst possible case, which is when the flow is brought to an immediate standstill by the instantaneous shutting of the valve at the base of the penstock.

Example on turbulence losses

A 40m penstock takes a flow of 0.5 m^3/s. The first 30m are made of pipe diameter 50cm, and the last 10m is of diameter 40cm. There is a 45° bend of radius 1.5m in the top section. The intake is a protruding pipe and there is a gate valve at each end of the penstock. What is the total head loss due to turbulence?

Flow velocity in top pipe $V_1 = \frac{4Q}{\pi D^2}$

$$V_1 = \frac{4 \times 0.5}{3.14 \times 0.5^2}$$

$$= \underline{2.6 m/s}$$

Similarly, flow velocity in bottom pipe

$$V_2 = \frac{4 \times 0.5}{3.14 \times 0.4^2}$$

$$= \underline{4.0 m/s}$$

For the 45° bend:

$R/D = 3$, so $K_{bend} = 0.3$

For the contraction:

$D_1/D_2 = 1.25$, so $K_{contraction} = 0.15$

Also, $K_{intake} = 0.8$

and $K_{valve} = 0.1$ for both gate valves.

Therefore the head loss is:

$$h_t = \frac{V_1^2}{2g} \times (K_{intake} + K_{valve} + K_{bend}) + \frac{V_2^2}{2g} \times (K_{contraction} + K_{valve})$$

$$= \frac{2.6^2}{2g} \times (0.8 + 0.1 + 0.3) + \frac{4^2}{2g} \times (0.15 + 0.1)$$

$$= \underline{0.6 m}$$

Table 6.4 Turbulence losses in penstocks

Head loss coefficients for intakes (K_{intake})

Entrance profile:				
$K_{entrance}$	1.0	0.8	0.5	0.2

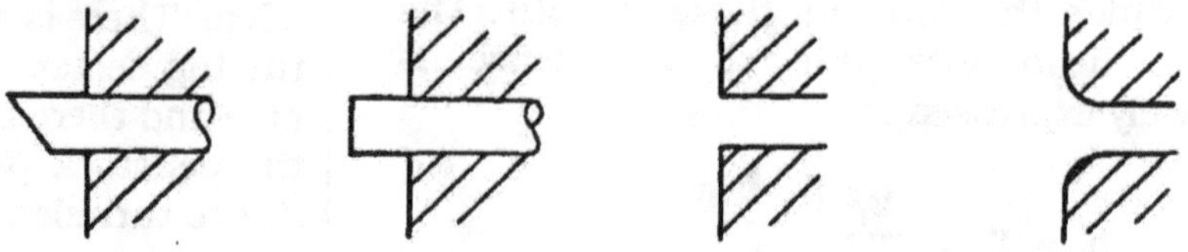

Head loss coefficients for bends (K_{bend})

Bend profile:

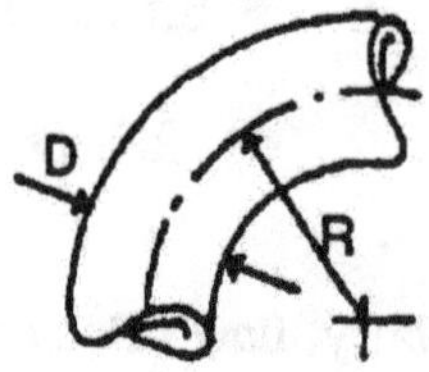

D

R

R/D		1	2	3	5
K_{bend}	θ = 20°:	0.36	0.25	0.20	0.15
	θ = 45°:	0.45	0.38	0.30	0.23
	θ = 90°:	0.60	0.50	0.40	0.30

Head loss coefficients for sudden contractions ($K_{contraction}$)

Contraction profile:

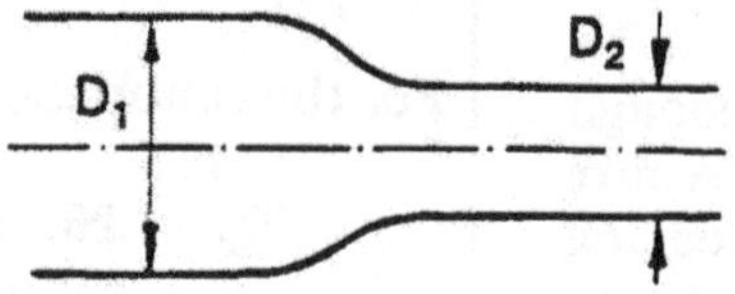

D_1 D_2

D_1/D_2	1.0	1.5	2.0	2.5	5.0
$K_{contraction}$	0	0.25	0.35	0.40	0.50

Head loss coefficients for sudden enlargements ($K_{enlarge}$)

D_1/D_2	0.2	0.4	0.6	0.8	1.0
$K_{enlarge}$	0.9	0.7	0.4	0.1	0

Head loss coefficients for valves (K_{valve})

Type of valve	Spherical	Gate	Butterfly
K_{valve}	0	0.1	0.3

Figure 6.9 Flowchart for calculating penstock pipe diameter

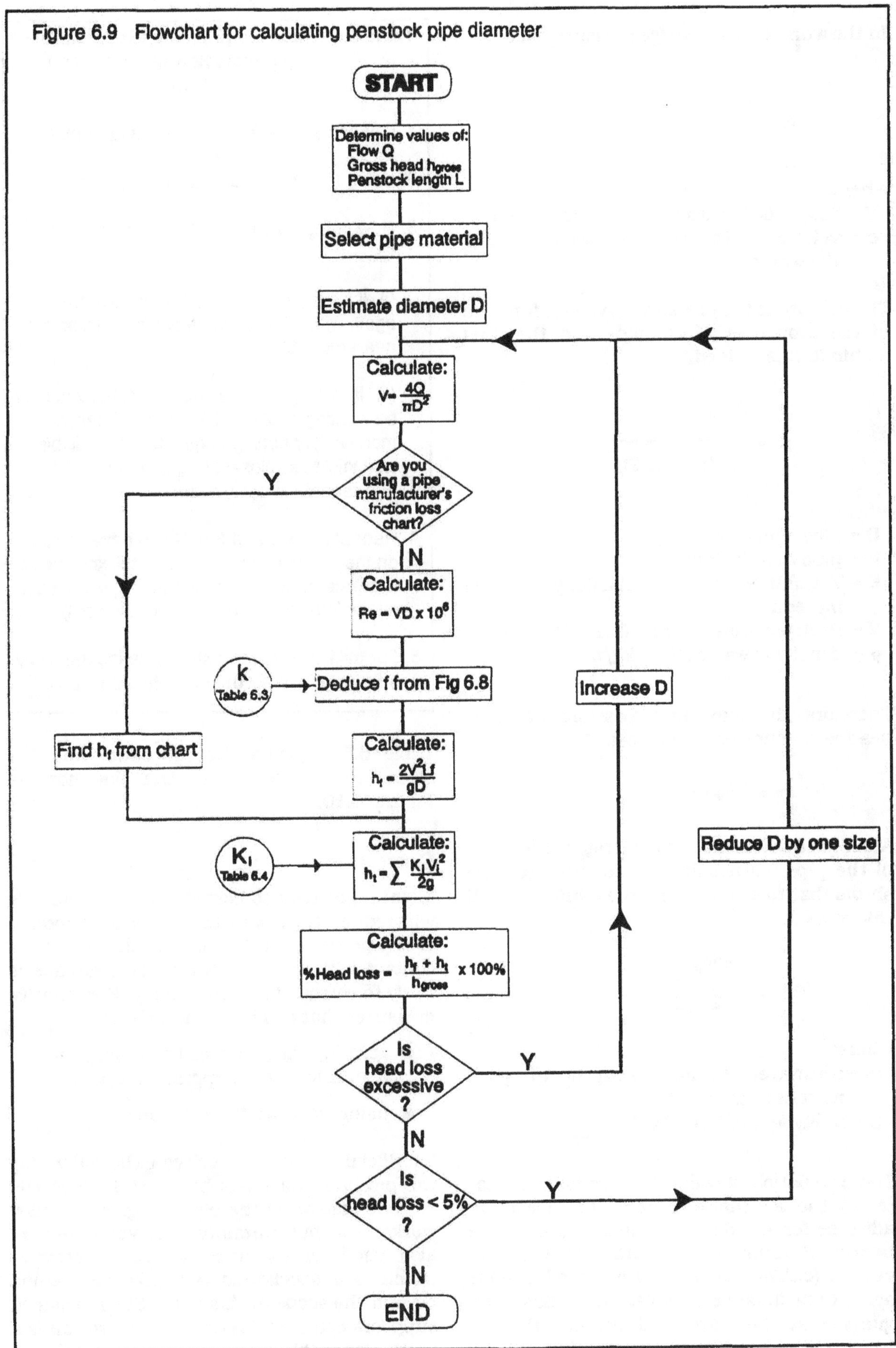

In the worst case the surge pressure head is:

$$h_s = \frac{c\,V}{g}$$

where:
V = flow velocity with valve fully open (m/s)
c = velocity of the pressure wave through the water (m/s)

The velocity of the pressure wave is a function of the properties of the pipe and the fluid within it. It is defined by:

$$c = \frac{1}{\sqrt{\rho(1/K + D/Et)}}$$

where :
D = pipe diameter
t = pipe wall thickness
E = Young's Modulus of Elasticity for pipe material
K = Bulk Modulus of water (2.1×10^9 N/m^2)
ρ = density of water (1000 kg/m^3)

Therefore the maximum possible pressure head experienced by the pipe is:

$$h_{max} = h_{gross} + h_s$$

A simple calculation of the stress in the walls of the pipe when subjected to this pressure shows that to avoid failure, the minimum wall thickness is:

$$t_{min} = \frac{\rho g h_{max} D}{2\sigma_T/S}$$

where:
σ_T = ultimate tensile strength of pipe material (Table 6.2)
S = safety factor, typically 3.

The calculation of wall thickness is complicated by the fact that a value for t is needed in advance for working out the surge pressure because it occurs in the equation for the wave velocity (c). The calculation therefore has once again to be iterative i.e. a value for t has to be guessed at the start, and the calculations have to be repeated to improve on this guess.

Table 6.5 Iterative procedure for calculating penstock wall thickness t. (see Figure 6.10)

1. Choose a value of t that seems appropriate.
2. Using this value, calculate c, h_{max} and t_{min}.
3. Now compare your chosen value of t with t_{min}:

   (i) If $t < t_{min}$ then the wall is too thin. Start again with a larger value of t; try a thickness nearer to $1.5t_{min}$.

   (ii) If $t > t_{min}$ then the wall is OK, but you may be wasting excess thickness (and hence money). Try putting t equal to the available wall thickness closest to t_{min}, above or below, and repeat the calculation.
4. Repeat steps 2 and 3 until you have settled on the minimum available wall thickness that is greater than t_{min}. (Note that t_{min} changes every time you choose a new value of t).
5. For mild steel pipes, t should be increased by 1.5mm to take account of corrosion.

Table 6.5 explains the procedure and the steps are illustrated in the flowchart of Figure 6.10.

Corrosion

Metal pipes tend to corrode in time, and this will reduce their wall thickness in places. If the pipe is buried underground or is in contact with wet soil, the outside surface is likely to corrode far more rapidly. Preventative measures that can be taken include:

- painting the penstock thoroughly and regularly with an appropriate paint
- using 'sacrificial' protection.

Sacrificial protection involves either (a) wiring the pipeline to a block of a metal which will corrode instead of the pipe; magnesium alloy works best, but aluminium alloys or zinc are also used, or (b) using *cathodic protection*, which is a specialized but effective method beyond the scope of this book. Steel penstock suppliers could advise on this if corrosion is a particular problem.

Summary

Table 6.6 is laid out to enable a methodical and error-free calculation of the diameter and wall thickness of the penstock, aided by the flowcharts in Figures 6.9 and 6.10. Successive iterations for D and t can be tabulated side-by-side. The table also enables the performances of different pipe materials to be compared directly with each other.

The aim should be to run through the upper half of the table enough times for an appropriate pipe diameter and associated head loss to be reached, and then to do the same in the lower half to arrive at a suitable pipe wall thickness.

It is a good idea to consult pipe suppliers while planning the penstock selection, so that all the pipe sizes which are available can be considered, and prices, delivery costs, etc. can all be taken into account. It is often found, for instance, that steel pipes are supplied with a standard wall thickness of either 4mm or 6mm.

It is recommended that PVC pipes are sized in the same way as other pipes. It is likely that the manufacturer of PVC pipe will provide information on both pressure rating and friction loss, but although these provide a quicker method of selecting PVC pipe, there is more flexibility in the choice of PVC pipe than suggested by a single manufacturer's pressure ratings. Also, the manufacturer is assuming some surge pressure in his ratings which may not be close to the surge in your system.

Figure 6.10 Flowchart for calculating penstock wall thickness

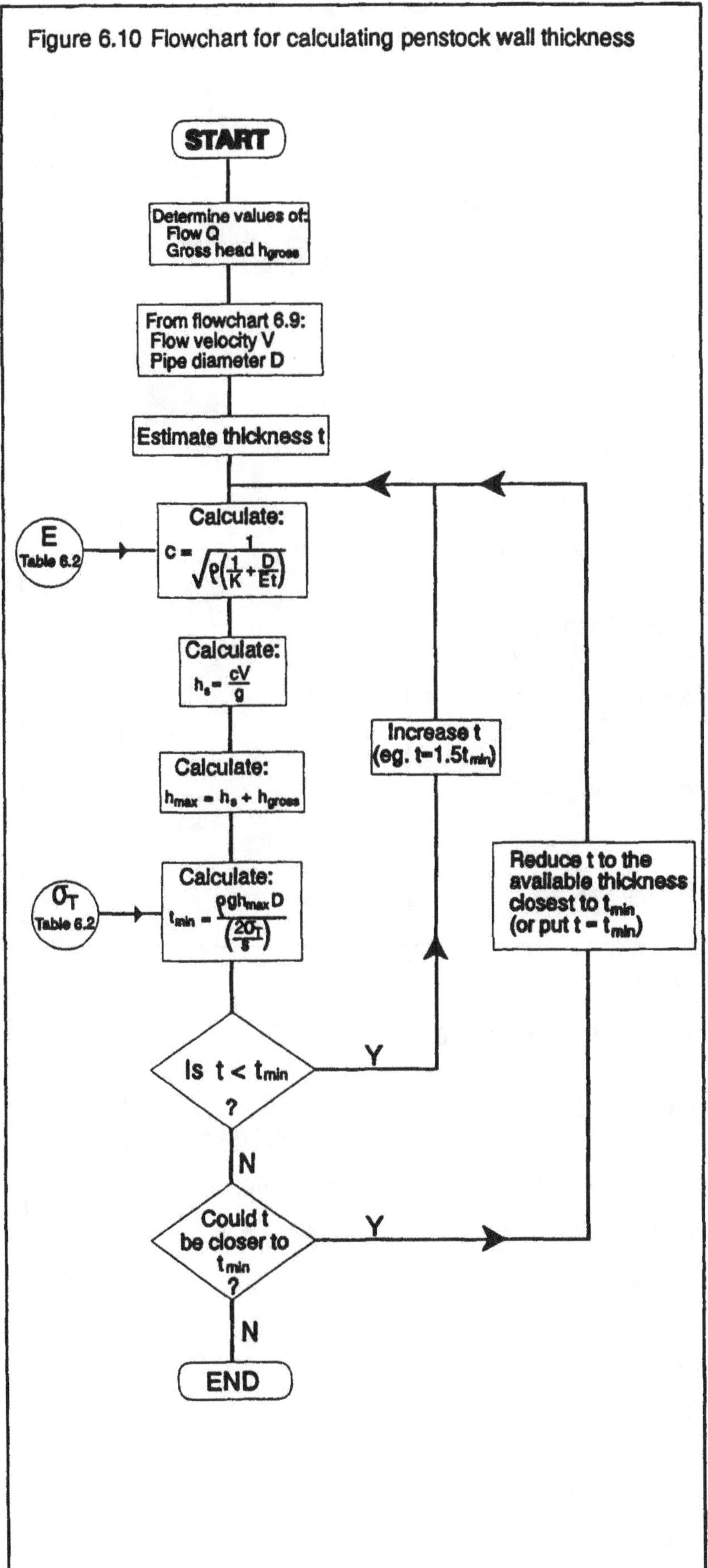

Table 6.6 Calculation sheet for penstock dimensions and cost

Constants:

Flow Q (in m^3/s) = Water Density ρ =1000 kg/m^3

Penstock length L (in m) = Bulk Modulus K = 2.1×10^9 N/m^2

Gross head h_{gross} (in m) = Gravity g = 9.8 m/s^2

Variable	Formula	Units	Pipe 1	Pipe 2
Material	-	-		
Diameter	D	m		
Flow velocity	$V = 4Q/\pi D^2$	m/s		
Reynolds Number	$Re = VD \times 10^6$	-		
Equiv. grain size (Table 6.3)	k	mm		
Friction factor (Figure 6.8)	f	-		
Friction head loss (or from Figure 6.7)	$h_f = \frac{2 V^2 L f}{g D}$	m		
Sum of turbulence loss. coefficients (Table6.4)	ΣK_i	-		
Turbulence loss	$h_t = \Sigma K_i \frac{V_i^2}{2g}$	m		
%Head loss	$100\% \times \frac{h_f + h_t}{h_{gross}}$	%		
Wall thickness	t	m		
Young's Modulus (Table 6.2)	E	N/m^2		
Pressure wave velocity	$c = \frac{1}{\sqrt{\rho(1/K + D/Et)}}$	m/s		
Max pressure surge	$h_s = cV/g$	m		
Total max head	$h_{max} = h_s + h_{gross}$	m		
Tensile strength (Table 6.2)	σ_T	N/m^2		
Safety factor	S	-		
Min. wall thickness	$t_{min} = \frac{\rho g h_{max} D}{2\sigma_T/S}$	m		
Price per metre	P	\$/m		
Total cost of pipe	P x L	\$		

6.7 Burying the Penstock

Penstock pipelines can either be surface-mounted or buried underground. The decision will depend upon the pipe material, the nature of the terrain, and environmental considerations.

PVC pipe

PVC pipe should generally be buried unless the penstock is quite short. This is for three reasons:

- if properly buried and backfilled it is evenly supported along its entire length, thus avoiding the stresses caused by having to bridge the supports which hold up a surface-mounted pipe;
- the pipe is not exposed to ultraviolet light which can degrade the surface of the PVC;
- temperature conditions are more stable underground, so the tendency of PVC pipes to shift due to expansion and contraction is reduced.

Steel pipe

Steel penstocks are usually better surface-mounted. They generally need regular repainting and are strong enough to bridge much more widely spaced supports without serious overloading. They are not affected by light and are less prone to expansion and contraction, especially if the line is permanently filled with flowing water. A buried steel penstock needs special attention because it is in a much more corrosive environment than an exposed pipeline, and is very difficult to inspect.

Laying the pipeline

The correct method for burying a PVC pipeline is indicated in Figure 6.11. Note that it should ideally be at least 750mm below ground level, especially if heavy vehicles such as tractors or trucks might drive over it. Where nothing heavy is likely to cross its path, a shallower trench will suffice.

Before laying the penstock, all sharp rocks and sizeable stones need to be removed from the trench. Additional excavation is usually needed at the points where lengths of pipe are to be connected. A bedding material may need to be laid along the bottom of the trench which should be completely free of large stones and preferably be granular (fine gravel graded to 5-10mm or coarse sand is ideal). Clay or other materials which cake when wetted are not suitable. The bedding material should be compacted by hand to ensure a level and even surface, and if bricks or other temporary supports have been used when installing the pipeline, scrupulous care should be taken to ensure they are removed before backfilling the trench.

If the trench passess through impermeable ground, it is wise to dig occasional drainage outlets to one side to divert any flow of water which could develop under the pipe and wash away the bedding.

After laying the pipe the trench should be backfilled. The initial backfill material should be compacted onto the pipe in 100mm layers, by hand. Any stones, roots or other debris must be removed. If necessary, suitably fine backfilling material should be imported from another site to cover the pipe to at least 100mm above its highest point. Old topsoil

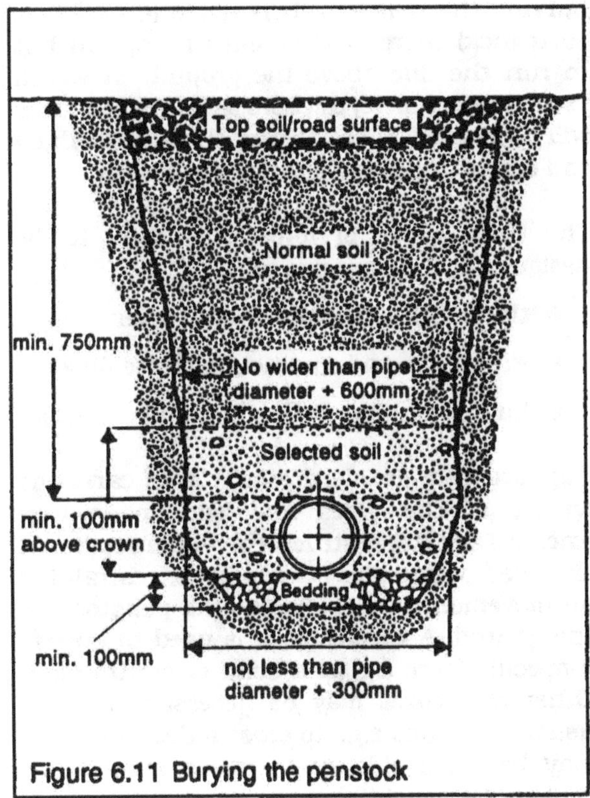

Figure 6.11 Burying the penstock

can then be used to fill the rest of the trench. Mechanical compactors should not be used on plastic pipelines because they may crack the PVC. Special care should be taken not to drive over the newly installed line with heavy equipment such as JCB excavators. If necessary such machines can be 'walked' across the pipeline using their hydraulic hoists. The buried pipeline ought to be clearly cordoned off while any heavy vehicles are on site.

A final consideration on whether to bury or surface-mount a penstock is that burying a pipeline removes perhaps the biggest eyesore of a hydro-scheme and greatly reduces its visual impact. However, out of sight is out of mind, and it is vital to ensure that a buried penstock is properly and meticulously installed because any subsequent problems, such as leaks, will be much harder to detect and rectify.

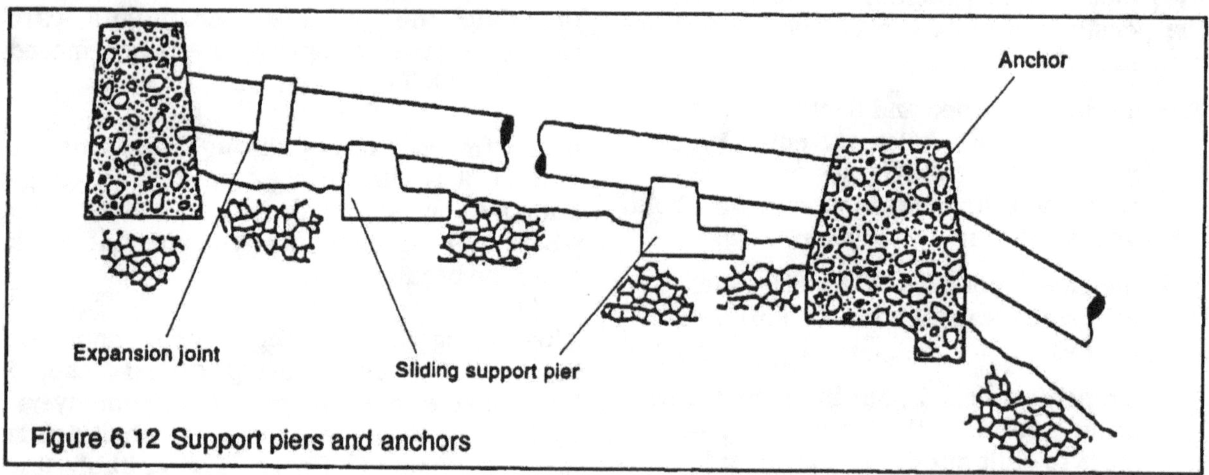

Figure 6.12 Support piers and anchors

6.8 Supporting the Penstock

Where the nature of the ground renders burying the penstock impossible (e.g. if it is solid rock) there is sometimes no option but to run the line above the ground, in which case support piers, anchors and thrust blocks will be needed to counteract the forces which can cause undesired pipeline movement.

The three types of force that need to be designed against are due to:

- the weight of the pipes plus water
- expansion and contraction of the pipe
- fluid pressures, both static and dynamic.

Support piers are used primarily to carry the weight of the pipes and enclosed water. Anchors are larger structures which represent the fixed points along a penstock, restraining all movements by anchoring the penstock to the ground. A thrust block is used to oppose a specific force e.g. at a bend or contraction. Other structures may be necessary for unusual situations e.g. to cross a deep ravine it may be more efficient to use a suspension cable or lattice girder.

The different support structures can usually be built of rubble masonry or plain concrete. Anchor blocks may need steel reinforcement, and triangulated steel frames are sometimes used for support piers. All supports should be placed on original soil and not fill. The bearing area must be calculated to support the pipeline without exceeding the safe bearing load of the soil. Drainage should be provided to prevent erosion of the support foundations.

The size and cost of the support structures for a given penstock are minimized by:

- keeping the penstock close to the ground
- avoiding tight bends
- avoiding soft/unstable ground.

These factors should be taken into consideration by the designer/contractor when choosing the line of the penstock. A further point to note is the importance of completing the work quickly so that the pipe can be filled with water as soon as possible. Temperature extremes caused by the sun shining on an empty pipeline tend to cause joints to drift apart due to excessive expansion or contraction.

Support piers

Support piers are positioned at regular intervals along a surface-mounted penstock. Their spacing depends on the strength of the penstock pipes and the joints. They must be able to accommodate longitudinal movement of the pipes due to expansion and contraction, either by allowing the pipes to slide over them, or by some hinging mechanism.

Wherever the penstock can be positioned close to ground level, simple supports built of rubble masonry or plain concrete can be used. Hinged support piers are suitable for carrying the penstock higher off the ground. They are pivoted at their base and move with the pipe as it expands and contracts.

It is usually not necessary to analyse each support pier in detail. The method suggested is to design one or two 'standard' piers which can be used along the whole penstock, provided certain conditions are satisfied.

Anchors

Anchors are simply blocks of reinforced concrete in which the penstock is embedded. Their main function is to fix the penstock securely in position and prevent it from sliding downhill. They are frequently placed at bends where they can oppose the hydrostatic force that pushes outwards as the flow changes direction. If expansion joints are not inserted between them, the anchors must also resist the thermally induced stresses which will attempt to lengthen or shorten the pipe as the temperature rises and falls.

Thrust blocks

These are concrete blocks which are usually inserted after the pipeline has been laid to oppose a specific force. They are required at bends where there is no anchor, as illustrated in Figure 6.13.

Support spacing

If in doubt, it is generally best to use one support for each length of pipe.

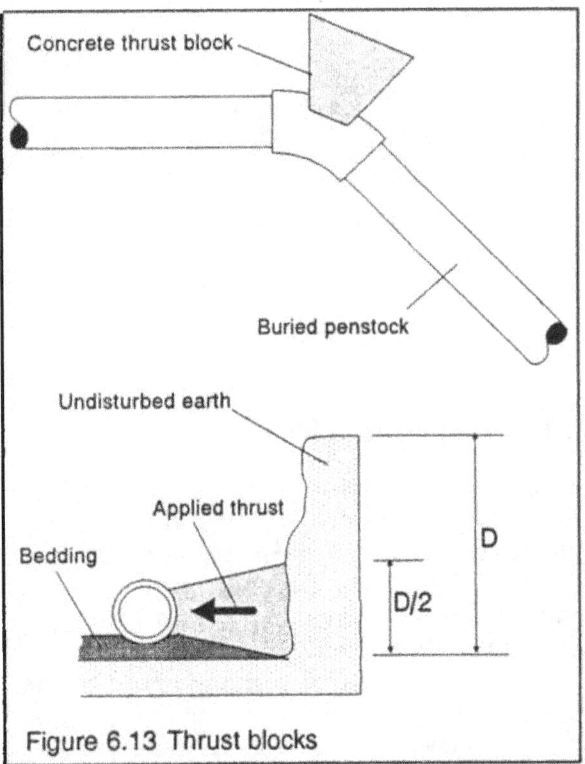

Figure 6.13 Thrust blocks

Steel pipes

The important criterion here is the jointing system. For any flexible coupling method, one support per length is needed. For flanges to British Standard (minimum thickness 16mm) and welds, the pipes can be considered as one piece; Table 6.7 is a guide to the necessary support-spacing, based on experience and pipe manufacturers' recommendations.

PVC pipes

Follow the pipe manufacturers' recommendations, which are almost always for one support per length, although burying the pipe is usually easier and less expensive with PVC and prevents damage by ultraviolet radiation.

Table 6.7 Spacing (in metres) of supports for welded or flanged steel pipes.

Wall Thickness (mm)	Pipe diameter (mm)				
	100	200	300	400	500
2	2	2	2.5	3	-
4	3	3	3	4	4
6	4	4.5	5	6	6

Turbines 7

7.1 Introduction

A turbine converts energy in the form of falling water into rotating shaft power. The selection of the best turbine for any particular hydro site depends upon the site characteristics, the dominant ones being the head and flow available. Selection also depends on the desired running speed of the generator or other device loading the turbine. Other considerations, such as whether the turbine will be expected to produce power under part-flow conditions, also play an important role in the selection. All turbines have a power-speed characteristic, and an efficiency-speed characteristic. They will tend to run most efficiently at a particular speed, head and flow.

Often the device which is driven by the turbine, usually an electrical generator, needs to be run at a speed greater than the optimum speed of a typical turbine. If so, speed-increasing gears (or pullies and belts) will be needed to link the turbine to the generator. It is preferable to minimize the speed-up ratio in order to reduce transmission costs and associated technical problems. (Transmission of power at low rpm involves heavier shafts, bearings and belts). As a rule of thumb, ratios of more than 3:1 should certainly be avoided, and less than 2.5:1 is preferable i.e. to operate a standard 1500rpm generator the selected turbine should run at designs speeds greater than 500rpm. In some cases it is possible to select a turbine which runs exactly at the required speed, so the generator can be coupled directly to the turbine. This is nearly always the case for large-scale hydro power, and manufacturers can supply complete turbine-generator sets. In micro-hydro installations it is usually cheaper to purchase separate units which must then be linked by a transmission drive.

A turbine's design speed is largely determined by the head under which it operates. Turbines can be crudely classified as high-head, medium-head, or low-head machines, as shown in Table 7.1. Turbines are also divided by their principal of operation and can be either *impulse* turbines or *reaction* turbines.

Table 7.1. Groups of impulse and reaction turbines

Turbine Type	Head: High	Head: Medium	Head: Low
Impulse	Pelton Turgo Multi-jet Pelton	Crossflow Turgo Multi-jet Pelton	Crossflow
Reaction		Francis	Propeller Kaplan

In Figure 7.1 the rotating element (or 'runner') of the reaction turbine is fully immersed in water and is enclosed in a pressure casing. The runner and casing are carefully engineered so that the clearance between them is minimized. The runner blades are profiled so that pressure differences across them impose lift forces, akin to those on aircraft wings, which cause the runner to rotate.

In contrast an impulse turbine runner operates in air, driven by a jet (or jets) of water, and the water remains at atmospheric pressure before and after making contact with the runner blades. In this case a nozzle converts the pressurized low velocity water into a high speed jet. The runner blades deflect the jet so as to maximize the change of momentum of the water, and hence maximize the force on the blades. The casing of an impulse turbine is primarily to control splashing because its interior is at atmospheric pressure. Impulse turbines are usually cheaper than reaction turbines because there is no need for a specialist pressure casing, nor for carefully engineered clearances, but they are also only suitable for relatively high heads.

Figure 7.1 The difference between impulse and reaction turbines

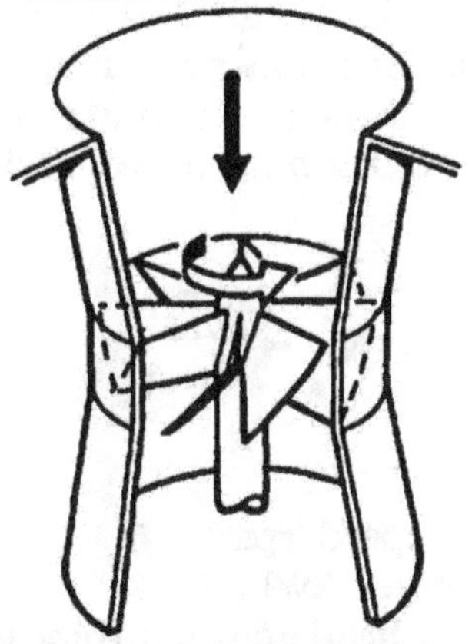

(a) Reaction turbines are turned by hydrodynamic lift forces acting on the runner blades

(b) Impulse turbines are turned by the momentum of a high speed water jet

Head-flow ranges

Each turbine type is best suited to a certain range of pressure head and flow rate. For instance, Pelton wheels operate with low flows discharged under great pressure, while at the opposite end of the spectrum, propeller turbines can accept huge volumes of water falling through only a few metres. The head-flow ranges applicable to each turbine type are summarized in the graph of Figure 7.2, which also indicates the power generated.

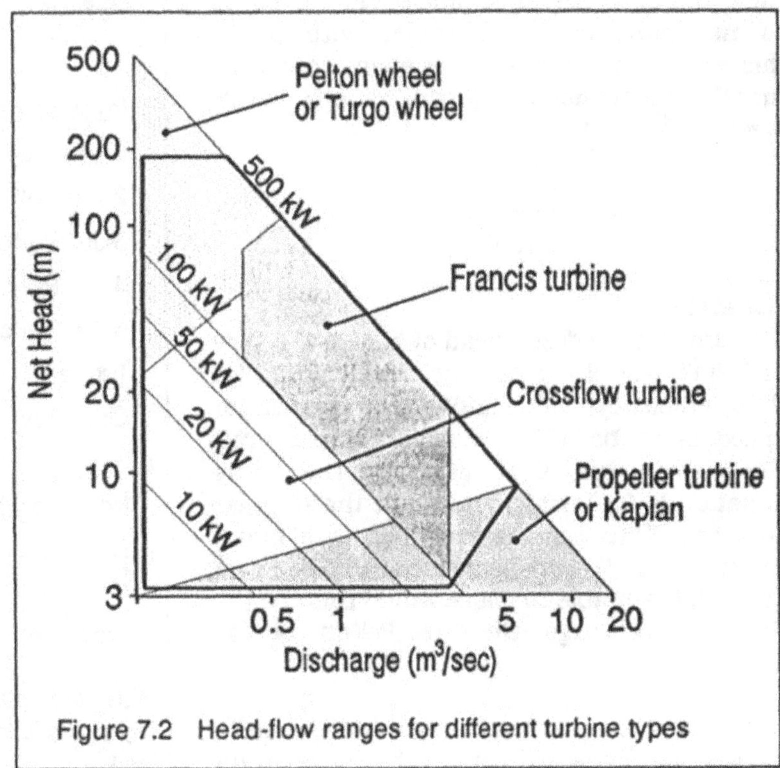

Figure 7.2 Head-flow ranges for different turbine types

Specific speed

If you are likely to be negotiating with turbine manufacturers, then it is important to know something about 'specific speed' (if not, you can skip this section).

Every turbine type has a numerical value associated with it called the *specific speed* which characterizes its performance. The specific speed relates the output power of the turbine to its running speed and the head across it.

$$\text{Specific speed } N_s = \frac{n_t P_o^{0.5}}{H^{1.25}}$$

where:
n_t =turbine speed (rpm)
P_o =shaft power (kW)
H =pressure head across turbine (m)

N_s does not depend on the *size* of the turbine. Identical turbines of different sizes have the same specific speed. Turbine manufacturers often give the specific speed of each of their machines in their literature. Typical values are listed in Table 7.2.

In principle the above equation enables the selection of a turbine which runs at exactly the required speed. If the head and output power are known (output power can be estimated from the input power by assuming a turbine efficiency of, say, 0.7) it is then possible to calculate the value of N_s which gives a turbine speed n_t equal to the alternator speed n_a. However in practice it is cheaper to use standard-size turbines with speed increasing transmission. The equation can be modified to include the gearing ratio G, where $n_t = n_a/G$, so that:

$$N_s = \frac{n_a P_o^{0.5}}{G\, H^{1.25}}$$

Example
A hydro site offering a head of 50m and a flow of 1.2m³/s implies a turbine output power (at 70% efficiency) of 412kW. The alternator speed is to be 1500rpm. With direct drive (G=1) the value of N_s given by the above equation is 230, which rules out the impulse turbines. With gearing stretched to a maximum of 3:1, N_s can be reduced to 75. From Table 7.2, suitable turbines would then be the Francis, the Turgo, the 6-jet Pelton, or the Crossflow.

Care should be taken with regard to the units of specific speed because the convention varies depending on the country. Table 7.2 is in metric units. A turbine classified, for instance, using the British Imperial system of units (ft, lbs, etc.) would have a specific speed 4.44 times smaller than that listed in Table 7.2.

Part-flow efficiency

Another significant factor in the comparison of different turbine types is their relative efficiencies at part-flow. Typical efficiency curves are shown in Figure 7.3.

An important point to note is that the Pelton, Crossflow and Kaplan turbines retain very high efficiencies when running below design flow; in contrast the efficiency of the Francis turbine falls away sharply if run at below half its normal flow, and fixed pitch propeller turbines perform very poorly except above 80% of full flow.

It may appear strange that the Francis turbine is as popular as it is, given that it tends to be a more complex and expensive machine, and has such a poor part-flow efficiency. The reason, as is clear from Table 7.2, is that the Francis is the only turbine suitable within a certain range of specific

Table 7.2 Specific speeds (metric) of various turbine types

Impulse turbines	Specific speed
Single-Jet Pelton	10 - 35
2-Jet Pelton	10 - 45
3-Jet Pelton	10 - 55
4-Jet Pelton	10 - 70
6-Jet Pelton	10 - 80
Turgo	20 - 80
Crossflow	20 - 90
Reaction turbines	
Francis	70 - 500
Kaplan	350 - 1100
Propeller	600 - 900

[N.B. the specific speed of a multi-jet turbine increases with the square root of the number of jets].

speeds. An impulse turbine operating under these conditions would be large, expensive, and cumbersomely slow-turning.

Suction head

Reaction turbines have high specific speeds which make them particularly suited to low-head applications. They also have a second important feature. As their method of power conversion is due to pressure differences across the blades, the low pressure below the blades (known as the *suction head*) is as effective in producing power as the positive pressure head above them. It is generally difficult or expensive to place a turbine lower than about 2m above the downstream water level. On a low-head site of, say, 10m, the suction head would then represent 20% of the power available at the site, which is likely to be very significant in terms of the overall economy of the scheme.

It is important to remember that the downstream water level will vary during the year. Therefore the bottom of the draft tube must extend to below the minimum water level expected, and the turbine must be situated above the maximum downstream flood level.

The casing which sustains the suction pressure downstream of the runner is known as the *draft tube*. Impulse turbines cannot normally make use of any reaction head because the casing runs at atmospheric pressure. There are exceptions to this rule where a vacuum is deliberately allowed to develop inside the casing to increase the total head.

Cavitation

The magnitude of the usable suction head is limited by an effect known as *cavitation*. This arises because very low water pressures are induced on the blades of a reaction turbine which is running under high suction. If the pressure falls below the vapour pressure of water at its current temperature, tiny bubbles of water vapour will form. The bubbles are carried by the flow to higher pressure regions where they suddenly collapse and give rise to shock waves which after a period of time will cause serious pitting and cracking of the blades.

Great care must be taken to situate the runner at a position which prevents the possibility of cavitation damage. (Information on setting the runner position is given at the

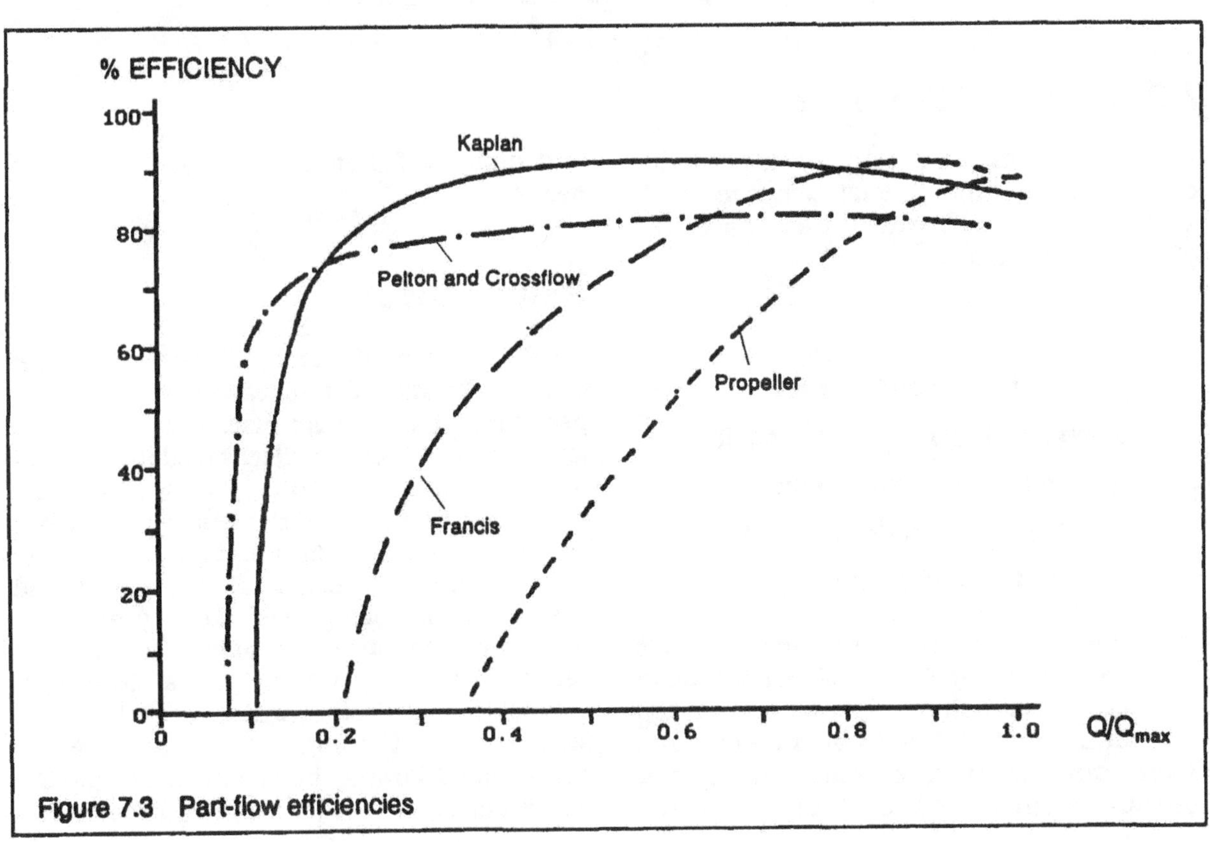

Figure 7.3 Part-flow efficiencies

end of Section 7.3.) The need to avoid cavitation often leads to the runner being set lower than ideally desired so that the suction head is reduced.

Runner size

In general smaller runners are cheaper because they use less material and rotate faster, so requiring less gearing. Casing costs will also tend to be less for a small diameter runner. In practice there are limits to how small the runner can be to handle the flow passing through it, but a well-designed turbine always has the smallest runner permissible within these limits.

Turbine costs

Typical costs for the different turbine types are given in Table 7.3. The variation with turbine size is also indicated. Costs are given as a range between the lowest and highest expected.

Table 7.3 Costs of turbines in units of US$1,000 (excluding alternater and drive)

Shaft Power (kW)	Crossflow	Francis	Single-jet Pelton	Multi-jet Pelton	Turgo	Propeller
2	1-2	4-6	2-4	1-3	2-4	4-6
5	2-6	8-10	3-8	3-6	5-8	8-10
10	2-10	15-20	5-15	4-10	8-14	15-20
20	3-14	20-30	8-20	6-15	12-20	20-30
50	5-30	25-70	20-50	15-30	35-50	25-70
100	30-50	40-100	40-80	30-60	55-80	40-100
150	50-80	60-120	60-100	45-80	80-100	60-120

7.2 Impulse Turbines

Impulse turbines are generally more suitable for micro-hydro applications compared with reaction turbines because they have the following advantages:

- greater tolerance of sand and other particles in the water
- better access to working parts
- no pressure seals around the shaft
- easier to fabricate and maintain
- not subject to cavitation
- better part-flow efficiency.

The major disadvantage of impulse turbines is that they are mostly unsuitable for low-head sites because of their low specific speeds; too great an increase in speed would be required of the transmission to enable coupling to a standard alternator. The Crossflow, Turgo and multi-jet Pelton are suitable at medium heads.

Pelton turbine

A Pelton turbine (Figure 7.4) consists of a set of specially shaped buckets mounted on the periphery of a circular disc. It is turned by jets of water which are discharged from one or more nozzles and strike the buckets. The characteristic profile of the Pelton bucket has evolved towards maximum efficiency through a combination of empirical experience and theoretical modelling over many years. The bucket is split into two halves so that the central area does not act as a dead spot incapable of deflecting water away from the oncoming jet. The cutaway on the lower lip allows the following bucket to move further before cutting off the jet propelling the bucket ahead of it, and also permits a smoother entrance of the bucket into the jet.

For maximum performance the jet should be deflected backwards by the bucket with nearly the same *relative* speed with which it strikes it. This implies maximum change of momentum. Since the bucket is travelling forwards, the jet velocity needs to be almost twice that of the bucket so that on being deflected back on itself it has virtually zero absolute velocity, having imparted most of its kinetic energy to the bucket. In fact the Pelton bucket is designed to deflect the jet through 165° (not 180°) which is the maximum angle possible without the return jet interfering with the following bucket or the oncoming jet.

In large-scale hydro installations Pelton turbines are normally only considered for heads above 150m, but for micro-hydro applications Pelton turbines can be used effectively at heads down to about 20m. Pelton turbines are not used at lower heads because their rotational speed becomes very slow and the runner required is very large and unwieldy. If runner size and low speed do not pose a problem for a particular installation, then a Pelton turbine can be used efficiently with fairly low heads. If a higher running speed and smaller runner are required then there are two further options:

- **Increasing the number of jets**
  Having two or more jets enables a smaller runner to be used for a given flow and increases the rotational speed. The required power can still be attained and the part-flow efficiency is especially good because the wheel can be run on a reduced number of jets with each jet in use still receiving the optimum flow.

- **Twin runners**
  Two runners can be placed on the same shaft either side by side or on opposite sides of the generator. This configuration is unusual and would only be used if the number of jets per runner had already been maximized, but it allows the use of smaller diameter and hence faster rotating runners.

Figure 7.4 Pelton wheels

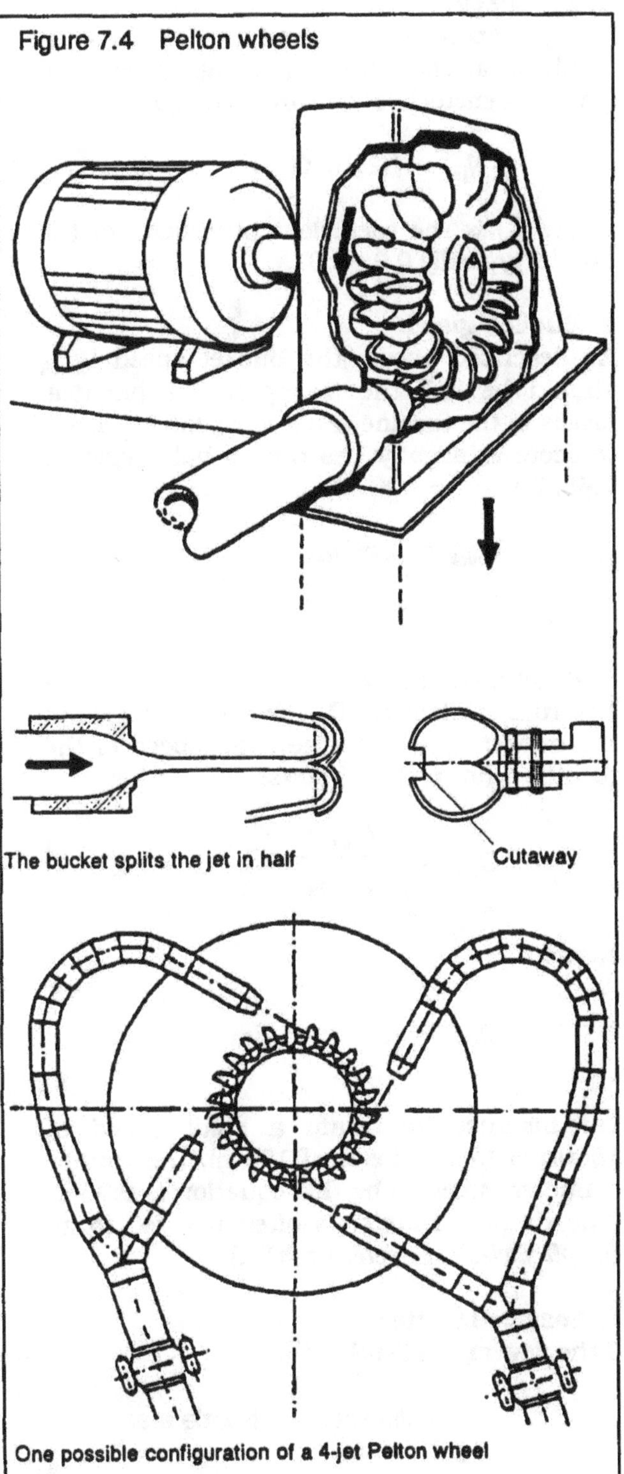

Theory

The equations detailed below can be used to provide quick estimates of the different geometries of Pelton wheel and nozzles which would be able to give the required output at a particular site. The four main parameters are: the runner diameter, the nozzle diameter, the number of buckets, and the number of nozzles.

1. Jet speed
The jet speed V_{jet} is fixed by the head H available at the nozzles (i.e. the gross head less the penstock losses) and is equal to:

$$V_{jet} = \sqrt{(2gH)}\ C_v$$

where C_v is the coefficient of velocity of the nozzle, typically 0.9 - 0.97.

2. Bucket speed
As described above, the bucket speed V_{buck} should be half the jet speed. In practice, losses in the turbine cause the peak efficiency to occur at slightly less than a half, typically 0.46. So:

$$V_{buck} = 0.46\ V_{jet}$$

$$= 2.0\sqrt{H}$$

3. Runner diameter
The runner diameter D_{run} is simply related to the bucket speed and the shaft speed of the turbine n_t (in rpm) as follows:

$$D_{run} = \frac{V_{buck}}{n_t \times \pi/60}$$

Hence:

$$D_{run} = 39\frac{\sqrt{H}}{n_t}$$

For instance, to attain a shaft speed of 1500rpm from a head of 100 m, the runner diameter indicated by this equation is 0.25m. (The runner diameter is often referred to as the *pitch circle diameter* or PCD.)

4. Nozzle diameter
If the flow rate is Q m^3/s then:

$$Q = \text{Jet speed} \times \text{Nozzle area}$$

$$= \sqrt{(2gH)} \times \frac{\pi D_{noz}^2}{4}$$

Rearranging,

$$D_{noz} = 0.54\frac{Q^{0.5}}{H^{0.25}}$$

So if the flow in the above example is 10*l*/s, the nozzle diameter works out to be 17mm.

5. Number of nozzles
If the flow in this example was 100*l*/s instead, then the nozzle diameter would need to be 54mm. A jet of this size would require buckets so large that only a few could be fitted around the runner of diameter 0.25m. There is therefore a second constraint on the size of the runner, which is that it must carry enough buckets for efficient operation, too few tending to waste water, too many causing interference. To match this constraint, experience has indicated that an acceptable ratio of runner diameter to nozzle diameter is 10, although Peltons can be made successfully with ratios in the range 6 to 20.

Rather than increase the runner diameter to meet the jet size, the other option is to spread the flow by using a number of jets. If the number of nozzles is N_{noz}, then the equation for the nozzle diameter becomes:

$$D_{noz} = 0.54\frac{(Q/N_{noz})^{0.5}}{H^{0.25}}$$

Fitted with 2 jets the turbine considered above could have nozzle diameters of 38mm, but this is still a bit too large for the runner. Fitted with 4 jets, the nozzle diameters are 27mm, and this is now suitable for the 25cm runner rotating at 1500rpm.

The maximum number of jets that can usually be arranged around a horizontal axis runner is 2, but a runner on a vertical axis can be fitted with up to 6 jets without them interfering with one another.

6. Number of buckets
The number of buckets N_{buck} required for efficient operation is generally found to follow the formula below:

$$N_{buck} = 0.5\frac{D_{run}}{D_{noz}} + 15$$

So our 4-jet wheel would therefore need 20 buckets.

7. Bucket width
For efficient operation, the bucket width should be at least 3 times the jet diameter.

Flow control and governing

Pelton wheels invariably drive electrical generators which must run at a fixed speed to keep the electrical output at a constant frequency and voltage. The bucket speed of the turbine must therefore not be allowed to vary. The speed of the turbine will increase as the load decreases, and vice-versa. To keep the speed constant, the input power to the turbine must be matched to the output power. One way of doing this is to control the water flow to the runner by varying the nozzle diameter with a *spear valve*.

Spear valves

A spear valve (Figure 7.5a) is so called because a streamlined spear-head is arranged to move within the nozzle to restrict the flow. It is able to vary the effective orifice area without creating turbulence. In doing so it alters the flow rate without affecting the velocity of the jet. If the loads are small, the spear valve can be closed to pass low flow. This means that in the case of hydro schemes which depend on water stored in a reservoir, less water is used when loads are small, and water is conserved for later use. The drawback of the spear valve is that it can only be operated slowly, since rapid closure gives rise to transient (or 'surge') pressures in the penstock which would require more robust and expensive pipe to be used. It therefore cannot be used for fast regulation of the flow. A further disadvantage is that it is a complex precision engineered component which can cost more than the turbine itself.

Deflector plates

A deflector plate (Figure 7.5b) rotates into the path of the jet so that the jet is partially or fully deflected away from the buckets. It reduces or totally diverts the flow reaching the buckets but does not alter the flow in the nozzle, nor the penstock and tailwater flow. The deflector plate creates no dynamic pressure variation and can be used instantaneously to no ill effect. Its disadvantage is that it does not conserve water, although this is not a problem in 'run-of-river' hydro schemes.

Deflector plates are not generally used to control speed but for emergency shut-down of the turbine.

Figure 7.5 Flow control

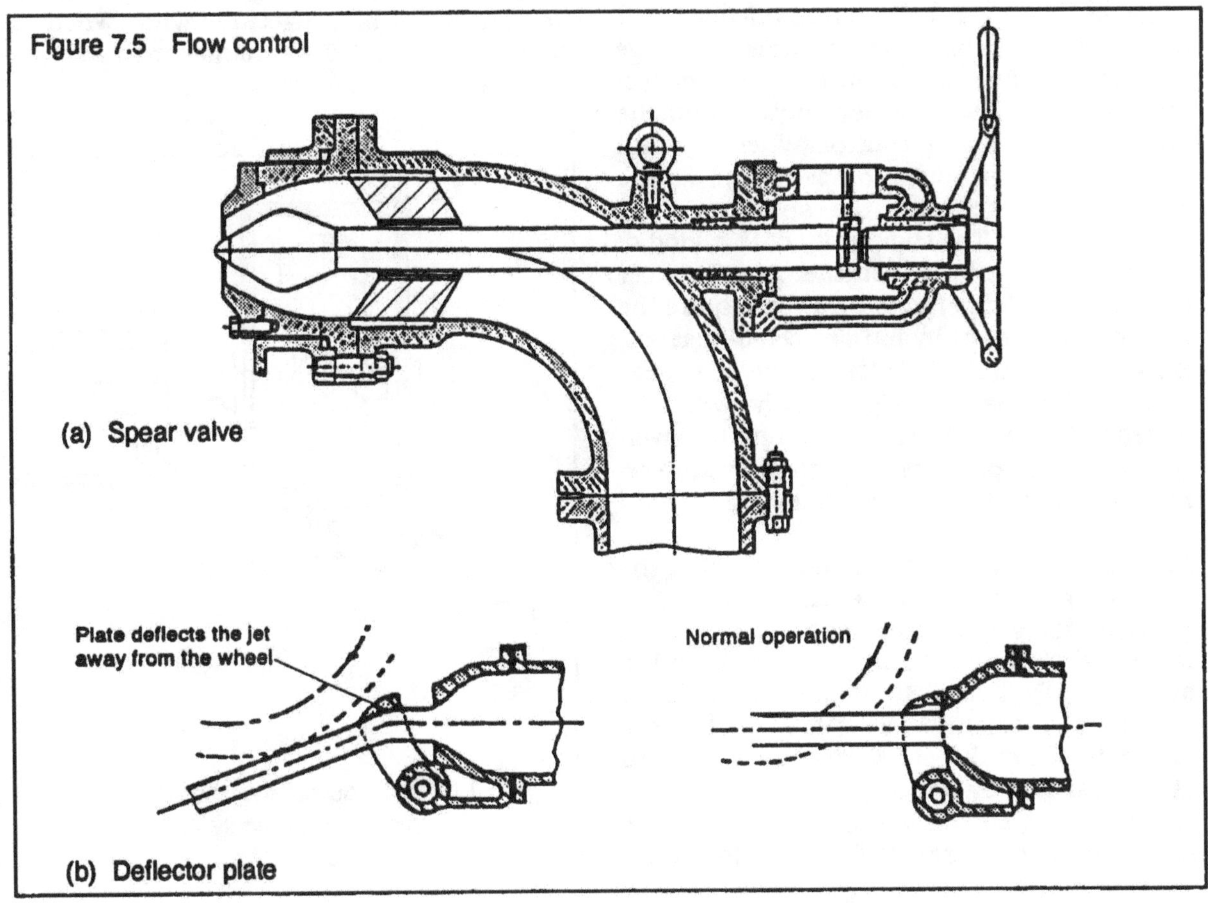

(a) Spear valve

(b) Deflector plate

Penstock valve

It is usual for the penstock to be fitted with a gate valve upstream of the turbine. This is to allow the operator to stop the penstock flow altogether so that the turbine can be isolated for maintenance. Gate valves must not be used as flow control devices by partially closing them, because the excessive turbulence created around the lip of the gate tends to produce cavitation effects which will quickly cause damage.

Speed governing

It is only recently that the multi-jet Pelton has been recognized as one of the most useful turbines available. The reason lies in the introduction of electronic load control methods for governing turbine speed. Speed governing is discussed fully in the next chapter; here it is only necessary to appreciate that without electronic load control, the alternative method for maintaining turbine speed during fluctuations in electricity demand is to regulate the flow of water entering the turbine.

In a Pelton turbine flow control governing is achieved by linking the spear valve or the deflector plate (or both) mechanically to a speed-sensing device. The governor of a large turbine will often link both, using the deflector plate as a fast-response control, and the spear valve as a fine control. When demand falls, the turbine starts to speed up, so the speed sensor activates the spear valve to reduce the flow. The complexity of this kind of mechanical governor is considerable. On most machines the large forces involved require the use of expensive hydraulic servo-mechanisms. For a multi-jet Pelton, these complications and costs rise significantly with the increase in the number of components involved. With a spear valve and deflector plate on each jet, the cost would be prohibitive.

An electronic load controller ensures that the output power is always the same irrespective of fluctuations in demand. Any fall in demand is sensed electronically and a ballast load is automatically added to maintain the total load on the generator. This means that the flexibility of a multi-jet Pelton can be achieved at modest cost by using electronics to adjust the electrical output rather than expensive spear valves to control the flow of water.

Turgo turbine

The Turgo turbine is an impulse machine similar to a Pelton turbine (Figure 7.6) but which was designed to have a higher specific speed. In this case the jet is aimed to strike the plane of the runner at an angle (typically 20°). The water enters the runner on one side and exits on the other. Therefore the flow rate is not limited by the discharged fluid interfering with the incoming jet (as is the case with Pelton turbines). As a consequence, a Turgo turbine can have a smaller diameter runner than a Pelton for an equivalent power. With smaller faster spinning runners, it is more likely to be possible to connect Turgo turbines directly to the generator, rather than having to go via a costly speed-increasing transmission.

Like the Pelton, the Turgo is efficient over a wide range of speeds and shares the general characteristics of impulse turbines listed for the Pelton, including the fact that it can be mounted either horizontally or vertically. A Turgo runner is more difficult to make than a Pelton and the vanes of the runner are more fragile than Pelton buckets. At one time they were exclusively made by Gilbert, Gilkes and Gordon, a UK manufacturer who owned the patent rights, but they are now manufactured in several other countries.

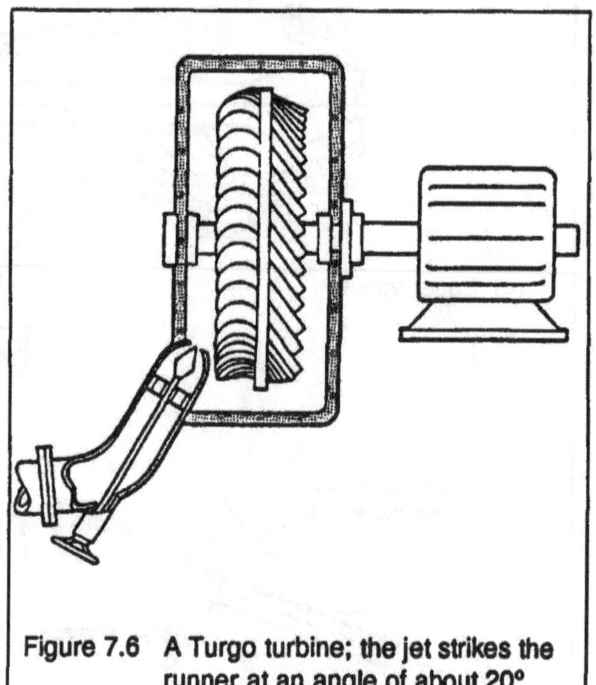

Figure 7.6 A Turgo turbine; the jet strikes the runner at an angle of about 20°

Crossflow turbine

Also called a Michell-Banki turbine, a Crossflow turbine has a drum-shaped runner consisting of two parallel discs connected together near their rims by a series of curved blades (Figure 7.7). A Crossflow turbine always has its runner shaft horizontal (unlike Pelton and Turgo turbines which can have either horizontal or vertical shaft orientation).

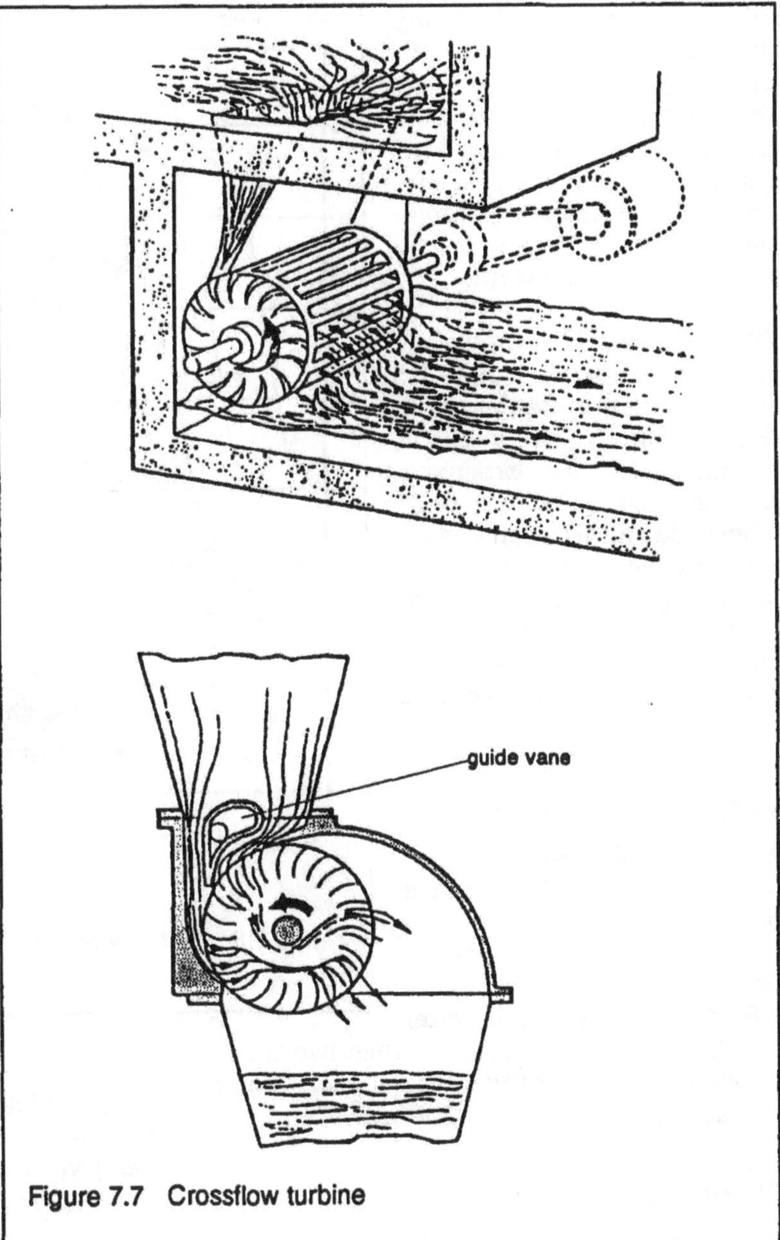

Figure 7.7 Crossflow turbine

Operation

In operation a rectangular nozzle directs the jet onto the full length of the runner. The water strikes the blades and imparts most of its kinetic energy. It then passes through the runner and strikes the blades again on exit, imparting a smaller amount of energy before leaving the turbine. Although strictly classed as an impulse turbine, hydrodynamic pressure forces are also involved, and a 'mixed flow' definition would be more accurate.

The intake to the turbine is fitted with a guide vane which is opened or closed by manual, hydraulic, or electrical means to regulate the flow to the turbine in the same way as a spear valve in a Pelton nozzle. The effective head driving the Crossflow runner can be increased by inducing a partial vacuum inside the casing. This is done by fitting a draft tube below the runner which remains full of tailwater at all times. Any decrease in the level creates a greater vacuum which is limited by an air-bleed valve in the casing. Careful design of the valve and casing is necessary to avoid conditions where water might back up and submerge the runner. This principle is in fact applicable to other impulse-type turbines but is not used in practice on any other than the Crossflow. It has the additional advantage of reducing the spray around the bearings by tending to suck air into the machine.

Due to the symmetry of a Crossflow turbine the length of the runner can theoretically be increased to any value without changing the hydraulic characteristics of the turbine. Hence, doubling the runner length merely doubles the power output at the same speed.

At high heads the Crossflow runner tends to be compact. The lower the head, the longer the runner becomes for a given power output. There are practical limits to length in both cases. If the blades are too long they will flex and fail due to fatigue at the junction of blade and disc. Intermediate bracing discs placed along the length of the runner can prevent this, but they reduce turbine efficiency by interfering with the water flow. In the case of a short runner operating at high head, efficiency losses at the edges become significant.

The efficiency of a Crossflow turbine depends on the sophistication of its design. A feature such as vacuum enhancement is expensive because it requires air seals where the runner shaft passes through the casing, as well as an air-tight casing. The most sophisticated designs attain efficiencies as high as 85%, the simpler ones achieve 65% - 80%.

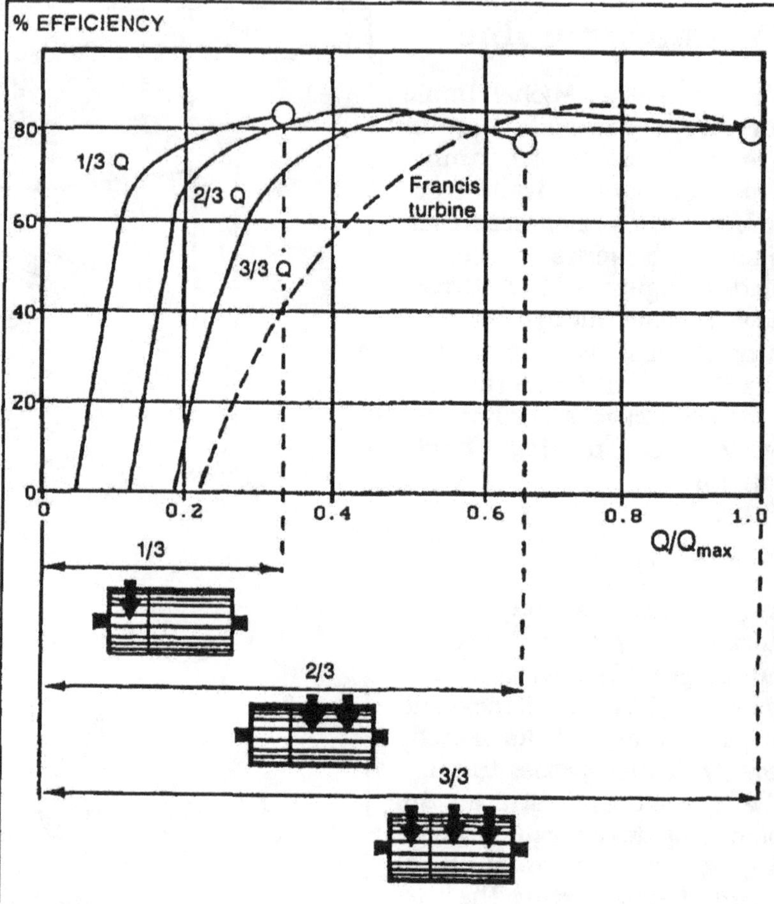

Figure 7.8 Part-flow efficiency of a partitioned Crossflow turbine

Part-flow efficiency

A high part-flow efficiency can be maintained at less than a quarter of full flow by the arrangement for flow partitioning illustrated in Figure 7.8. At low flows, the water can be channelled through either two-thirds or one-third of the runner, thereby sustaining a relatively high turbine efficiency.

Sizing

For sizing calculations on the Crossflow the reader is referred to Arta and Meier's paper in the bibliography. The dimensions of interest are the runner length L_{run} and diameter D_{run}, and the jet thickness t_{jet}. The length of the rectangular jet orifice is always equal to the runner length, while the jet thickness is designed for optimum performance.

If n_t is the required runner speed in rpm, Q is the flow in m³/sec, and H is the head in metres, then first approximations for the physical dimensions are given by the following equations:

$$D_{run} = 40 \frac{\sqrt{H}}{n_t}$$

The jet thickness is usually about one-tenth the runner diameter and the runner length is then given by:

$$L_{run} = \frac{0.23\,QH}{t_{jet}}$$

In summary, two major attractions of the Crossflow have led to considerable interest in this type of turbine:

- it is a design suitable for a very wide range of heads and power ratings (see Figure 7.2).
- compared to the other common types of turbine, it lends itself to particularly simple fabrication techniques, a feature which is of particular importance to developing countries. The runner blades, for instance, can be fabricated from lengths of pipe cut into strips.

7.3 Reaction Turbines

The reaction turbines considered here are the Francis turbine and the propeller turbine. A special case of the propeller turbine is the Kaplan. In all these cases, specific speed is high, i.e. reaction turbines rotate faster than impulse turbines given the same head and flow conditions. This has the very important consequence that a reaction turbine can often be coupled directly to an alternator without requiring a speed-increasing drive system. Some manufacturers make combined turbine-generator sets of this sort. Significant cost savings are made in eliminating the drive, and the maintenance of the hydro unit is very much simpler. Figure 7.2 indicates that the Francis turbine is suitable for medium heads, while the propeller turbine is more suitable for low heads.

On the whole, reaction turbines require more sophisticated fabrication than impulse turbines because they involve the use of larger and more intricately profiled blades together with carefully profiled casings. The extra expense involved is offset by high efficiency and the advantages of high running speeds at low heads from relatively compact machines.

Fabrication constraints make these turbines less attractive for use in micro-hydro in developing countries. Nevertheless, because of the importance of low-head micro-hydro, work is being undertaken to develop propeller machines which are simpler to construct. All reaction turbines are subject to the danger of cavitation, and most tend to have poor part-flow efficiency characteristics (Figure 7.3).

Figure 7.9 Francis turbine in a volute casing

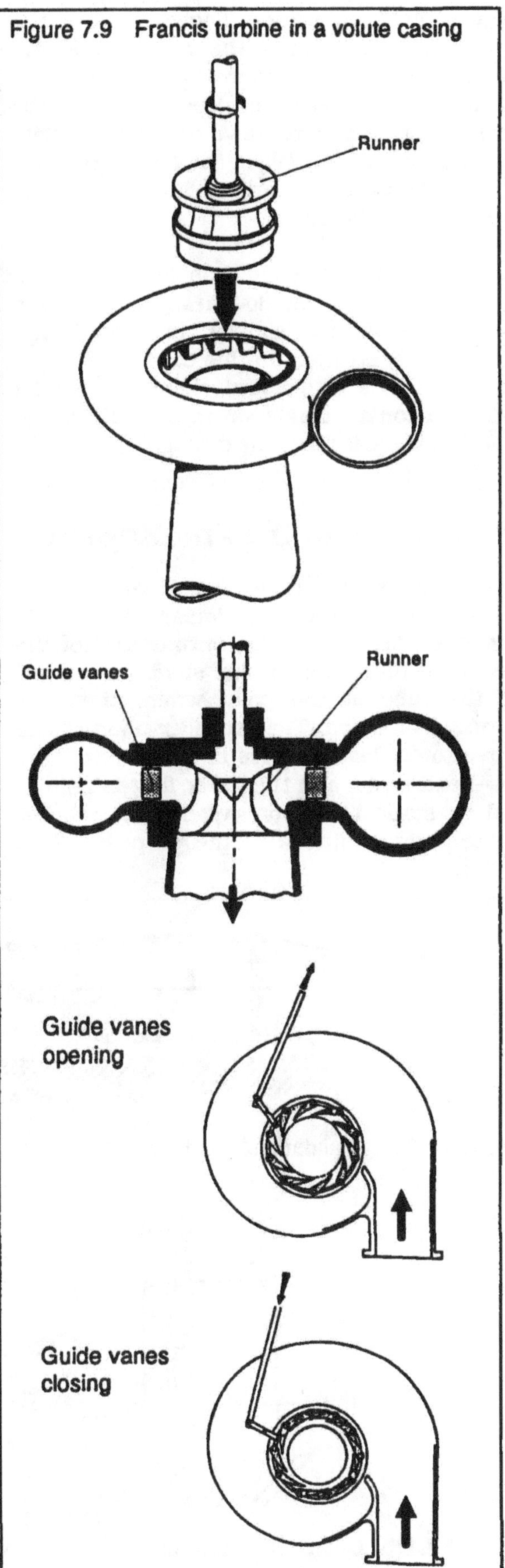

Francis turbine

Francis turbines can either be volute-cased or open-flume machines. Figure 7.9 illustrates an enclosed Francis turbine. The spiral casing is tapered to distribute water uniformly around the entire perimeter of the runner, and the guide vanes feed the water into the runner at the correct angle. The runner blades are profiled in a complex manner and direct the water so that it exits axially from the centre of the runner. In doing so the water imparts most of its 'pressure' energy to the runner before leaving the turbine via a draft tube.

The Francis turbine is generally fitted with adjustable guide vanes. These regulate the water flow as it enters the runner, and are usually linked to a governing system which matches flow to turbine loading, along the same lines as a spear valve or deflector plate in a Pelton turbine. When the flow is reduced the efficiency of the turbine falls away, as is evident from Figure 7.3.

The open-flume configuration in Figure 7.10 is often chosen for low-head applications because it is far simpler to install and significantly cheaper. In this case the turbine is placed in a water-filled chamber with its axis horizontal. Guide vanes still direct the flow but there is no volute casing.

Figure 7.10 Open-flume Francis turbine

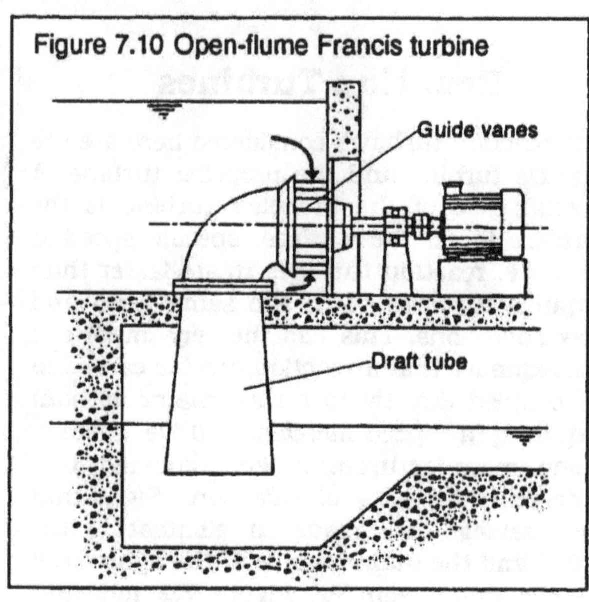

Propeller turbine and Kaplan

The basic propeller turbine (Figure 7.11a) consists of a propeller, similar to a ship's propeller, fitted inside a continuation of the penstock tube. The turbine shaft passes out of the tube at the point where the tube changes direction. The propeller usually has three to six blades, three in the case of very low-head units, and the water flow is regulated by static blades or swivel gates ('wicket gates') just upstream of the propeller. This kind of propeller turbine is known as a *fixed blade axial flow* turbine, because the pitch angle of the rotor blades cannot be changed. Although traditionally the propeller blade is profiled to optimize the effect of lift forces acting on it, designs have been produced with simple parallel section 'blunt' blades which offer less efficiency but are more easily fabricated. These are particularly applicable for micro-hydro where low cost and ease of

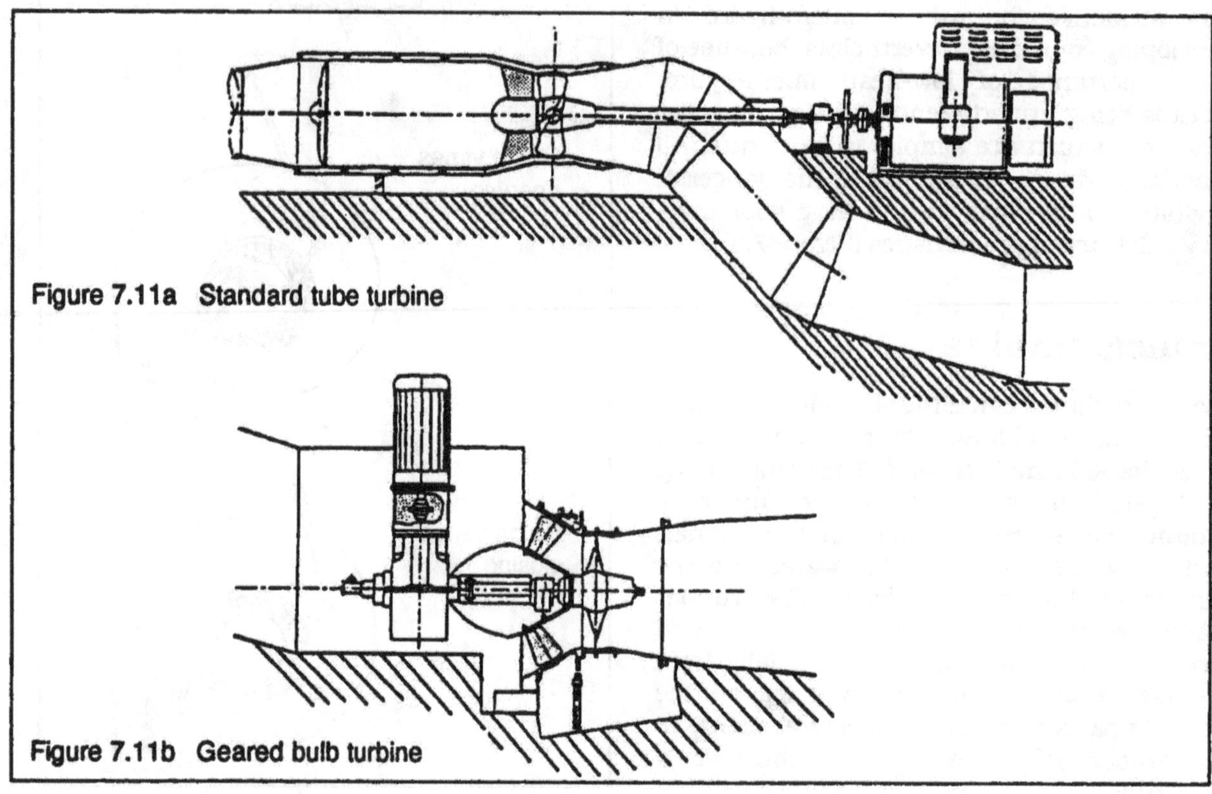

Figure 7.11a Standard tube turbine

Figure 7.11b Geared bulb turbine

fabrication are priorities. The part-flow efficiency of fixed-blade propeller turbines tends to be very poor.

Three common configurations for installing propeller turbines in micro-hydro installations are described below.

1. The runner is placed in a bend in the water passageway so that the shaft can protrude through the wall (upstream or downstream of the runner) to an externally mounted generator. This is sometimes known as a *tube turbine* (Figure 7.11a).
2. Known as a *geared bulb turbine* (Figure 7.11b), it has a right-angle drive, (using bevel gears) within the expanded hub of the turbine which allows the alternator drive-shaft to exit vertically from the tube, therefore avoiding the need for a bend.
3. The more sophisticated ungeared version of the bulb turbine incorporates the generator inside a casing (the 'bulb') which is an extension of the propeller hub, either upstream or downstream of the runner. This arrangement eliminates the need for the shaft to pass through the wall of the water passageway and is very compact. Bulb turbines also reduce the need for excavation at very low-head sites and there is no need for a powerhouse to protect the alternator. However, the special alternators required tend to make bulb turbines relatively expensive.

Kaplan

Large-scale hydro sites make use of more sophisticated versions of the propeller turbine. Varying the pitch of the propeller blades together with wicket gate adjustment enables reasonable efficiency to be maintained under part-flow conditions. Such turbines are known as *variable-pitch* or *Kaplan* turbines.

The wicket gates are carefully profiled to induce tangential velocity in the water (known as *whirl*). Common designs include the use of a scroll casing (similar to the Francis turbine) which feeds the water radially inwards, as illustrated in Figure 7.12. Whether radial or axial flow, variable pitch designs involve com-

Figure 7.12 Variable-pitch propeller or Kaplan turbine

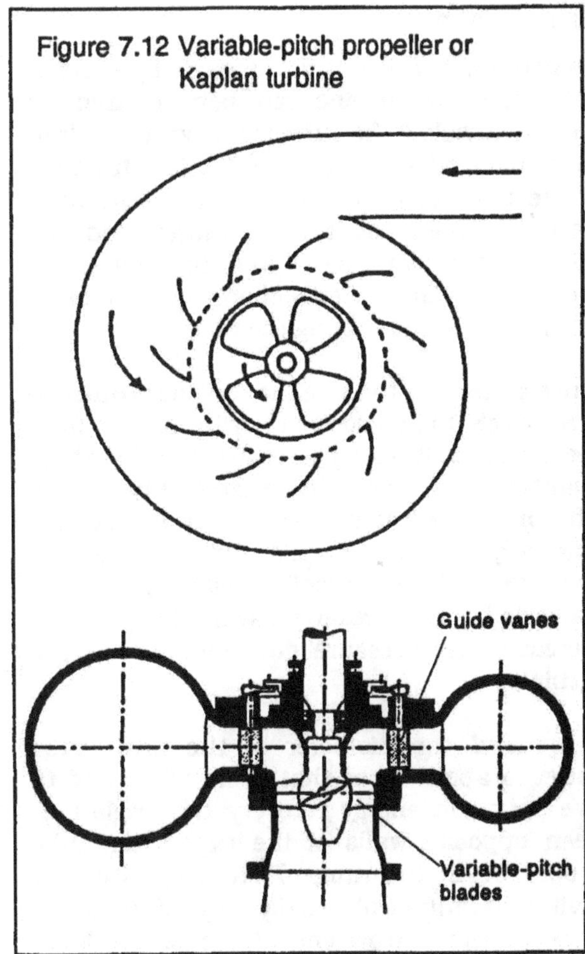

plex linkages which are disproportionately expensive for small systems and are usually not cost-effective in any except the largest of micro-hydro applications.

Reverse pump turbines

Centifugal pumps can be used as turbines by passing water through them in reverse. Research is currently being done to enable the performance of 'pumps as turbines' to be predicted more accurately.

The potential advantages are the low cost due to mass production (and in many cases also local production), the availability of spare parts, and the wider dealer/support networks. The disadvantages are the as yet poorly understood performance characteristics, lower efficiency, unknown wear characteristics, and very poor part-flow efficiency. Various companies have used pumps as turbines at various times, but the technology remains unproven and relatively poor in performance.

Draft tubes

All reaction turbines and the partially evacuated Crossflow turbine can benefit from an enclosure below the runner known as a draft tube, illustrated in Figure 7.13. The function of the draft tube is to maintain a column of water between the turbine outlet and the downstream water level and to enable the recovery of the kinetic energy (the *velocity head*) of the water leaving the runner.

Since water has to leave the turbine runner at a relatively high velocity in order to exit from the turbine, it still possesses a substantial quantity of energy. To recover this energy efficiently, the water velocity must be reduced gradually by a widening out of the exit passageway. The velocity head is thereby converted into suction pressure head which increases the pressure difference across the turbine.

In general the diameter of the draft tube outlet is about twice that of its inlet, and to give optimum energy recovery the angle between opposite walls of the expanding tube should be in the range 7° to 20°, although cavitation limits will usually require the angle to be at the narrow end of the range. Most manufacturers either supply a correctly designed draft tube or give detailed specifications for the construction.

The draft tube generally contains a partial vacuum, so the outlet must remain well submerged below the water surface to prevent air being sucked into the tube and displacing the water column. The difference in level between runner exit and tailwater level is called the *static suction head*.

Cavitation limits

The use of draft tubes with reaction turbines requires a knowledge of cavitation limits. The turbine manufacturer will advise on the correct position of the runner above the tailwater level to avoid cavitation. Suitable texts on the subject (e.g. Fritz) are listed in the bibliography.

In summary, the maximum allowable distance in metres between the runner and the tailwater level Z_{max} is given by :

$$Z_{max} = H_{atm} - H_{vap} - S_c H_{net}$$

Figure 7.13 Elevation and plan of a typical draft tube

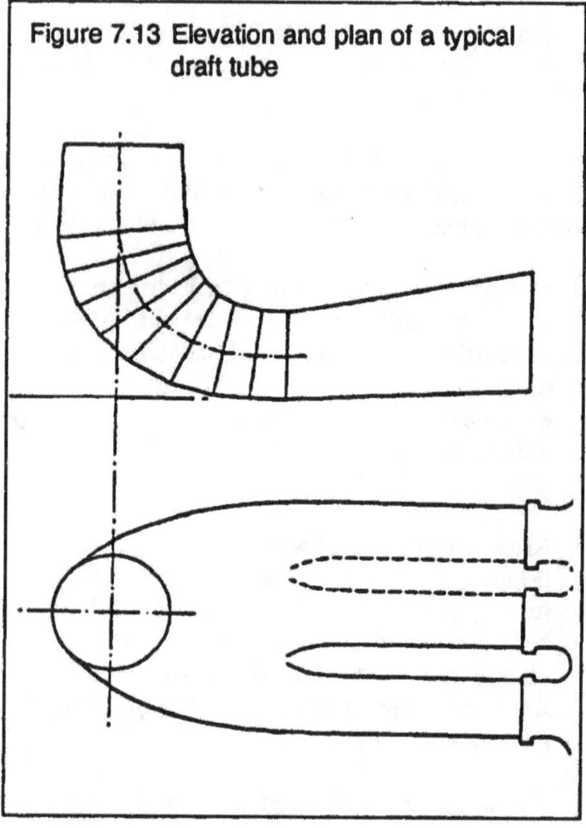

where:

H_{atm} = atmospheric pressure in metres of water

H_{vap} = the vapour pressure at the site water temperature

H_{net} = the net head at the turbine

S_c = a characteristic constant of the machine known as *Thoma's cavitation parameter*.

Like specific speed, S_c is a value provided by the manufacturer. Z_{max} is positive for a runner positioned above the tailwater level. The runner must be positioned below this value of Z_{max} if cavitation is to be avoided.

Governing

8

8.1 Introduction

A governor is a system which controls the speed of a turbine. A constant turbine speed is essential in electricity generating systems for maintaining a stable voltage and frequency output, but it is also generally required for most mechanical end-uses. Safety considerations usually demand a 'fail-safe' governing system because over-speeding or over-voltage can be dangerous to customers or users.

Generator speed

When electricity is generated at an isolated site by a synchronous generator, its frequency is determined by the speed of the generator and the number of poles. For example a four-pole generator produces two cycles per revolution of its shaft. To generate the standard 50cycles/sec (or 50Hz), it must therefore run at 25rev/sec, which is equivalent to 1500rpm. If this speed fluctuates then so does the frequency generated. Although most generators have some form of voltage regulation, the voltage output is also somewhat affected by speed changes.

Electrical equipment is designed to run at a specific voltage and frequency. Operation at frequencies or voltages other than the design values can seriously damage the appliances. For example, an electric motor will run hot if the frequency is too low, and it may burn out rather than start if the voltage is too low. If the voltage is too high most equipment will draw too much current and fail prematurely due to overheating.

Balancing input and output

Some control of generator speed is therefore essential. The speed of the turbine-generator set is determined by the input water power and the output electrical (or mechanical) power. If the input power matches the output power, then the system is stable. If the input is greater than the output, then the excess power will cause the turbine and generator to speed up. Conversely, if more power is being taken out then is being put in, then the turbine and generator will slow down. Governing therefore involves balancing the input power with the output power so as to keep the system running at a constant speed.

Flow control

Until recently, the usual method of governing was *flow control*, in which the flow of water through the turbine was adjusted so as to match the input water power with the demanded output power. This method can be summarized as follows.

When electrical power is required of the plant, the load on the generator increases and its speed starts to fall. Through either mechanical means (e.g. using a centrifugal flyball governor) or electrical means, the governor senses this speed reduction. It then opens the appropriate valves, often using hydraulic or electrical actuators to provide the necessary force, and more water is admitted to the turbine. Similarly, if less power is required, the governor senses an increase in turbine speed and reduces the flow through the turbine until the speed returns to normal.

Load control

Modern advances in electronics have now enabled an alternatve method of governing. Instead of adjusting the flow to match the fluctuating size of the load, it is now possible to control the magnitude of the load so that adjustments in the flow are no longer necessary. This method, known as *load control,* involves the deployment of a ballast load which soaks up extra power from the generator when the demand falls. In this way, the turbine can be allowed to run continuously at full power. Note that this method is less suitable when the supply of water is limited.

Where the turbine is directly linked to mechanical equipment, a governor is needed only if the running speed of the equipment is critical. For some mechanical uses, such as water pumping, this is not the case and governing is unnecessary.

8.2 Selecting a Governer to Suit the Load

In most cases it is best to purchase a turbine and generator set which includes an appropriate governing device supplied and guaranteed by the manufacturer. But before specifying the governing system, it is necessary to find out the tolerance of the end-use machinery to variations in frequency and voltage. Although most electrical machinery can put up with a ±10% fluctuation in voltage, the frequency must usually be held much closer to the nominal figure. Modern automatic voltage regulators (AVRs, covered in Section 10.5) will maintain a steady voltage over the acceptable range of frequencies.

Electrical loads

Heating

Heating loads are the most tolerant of variations in supply, in fact frequency variations do not affect these loads at all. Under-voltage simply reduces the heat output, but over-voltage can generate excess heat and cause the elements to burn out.

Lighting

Incandescent bulbs are not affected by frequency. Under-voltage decreases light output very sharply, but significantly increases bulb life, unless the voltage is fluctuating significantly. Over-voltage greatly reduces bulb life. For example, an over-voltage of only 5% reduces the life of bulbs by up to 50%. Quite small voltage variations can cause large variations in brightness.

Fluorescent lamps are affected by both voltage and frequency variations. If the voltage is more than 15% down, the lamp will not light. If the lamp is already operating, it will flicker more as voltage decreases. If the voltage drops by more than 25%, the lamp may go out. Small voltage variations only slightly affect the brightness.

Transformers

Transformer losses appear as heat. At fixed frequency, these losses vary approximately with the square of the voltage. Over-voltage can therefore pose a problem, and voltage is usually only allowed to increase by 5% at rated load. Under-voltage has no adverse effects. At fixed voltage, decreases in frequency lead to increased losses and heat generation.

Motors

Motors and transformers are affected in similar ways. Under-frequency with steady voltage causes high currents and overheating, and over-voltage at constant frequency has a similar effect. Manufacturers often specify that motors can operate satisfactorily at voltages within 10% of their rated value. Small motors with wide-range voltage regulators have been used successfully on ungoverned machines, but the motor life is reduced and this arrangement cannot be widely recommended.

8.3 Methods of Governing

The two general methods of governing fall under the headings of:

- flow control

or

- load control

Flow control governors

Conventional approaches to governing originated in the Eighteenth century during the early Industrial Revolution. The technology is largely based on mechanical and hydraulic principles but an increasing number of the components are now being replaced by electronic parts.

Hydraulic flow control

The general principle is that oil pressurized by a pump is used to drive a servomotor piston which moves the flow-control mechanism. Although hydraulic governors are used extensively on large hydro projects, they are sophisticated devices and their cost does not decrease in proportion to the size of the turbine. They therefore tend to be excessively costly for small-scale systems. They also require regular maintenance and are difficult to repair.

Mechanically driven

Typically, mechanical governors incorporate a flyball arrangement driven by the turbine shaft which effectively measures turbine speed. When an increased load is placed on the turbine, the resulting reduction in speed of the flyball assembly causes the flyballs to drop. As Figure 8.1 illustrates, this motion levers open a valve and allows oil under pressure into the upper chamber of a servomotor. The oil pressure moves the servomotor piston with great force and opens up the turbine valve (e.g. a spear valve for a Pelton wheel), thus allowing more water in through the turbine. The greater flow allows the turbine to generate the extra power required to meet the new load.

Mechanical feedback is provided by a dashpot and spring which connects the piston to the flyball lever. This prevents the governor from over-compensating and brings overall stability to the system. A manual control on the governor is used to set the operating speed and to start or shut down the system.

Electronically driven

Electro-hydraulic governors sense the turbine speed by electronically monitoring the output frequency. The frequency measurement is compared continuously with a reference value and the difference results in electronic commands that are transmitted to the hydraulic part of the governor by electrical actuators.

Figure 8.1 Hydraulic flow control with flyball governer

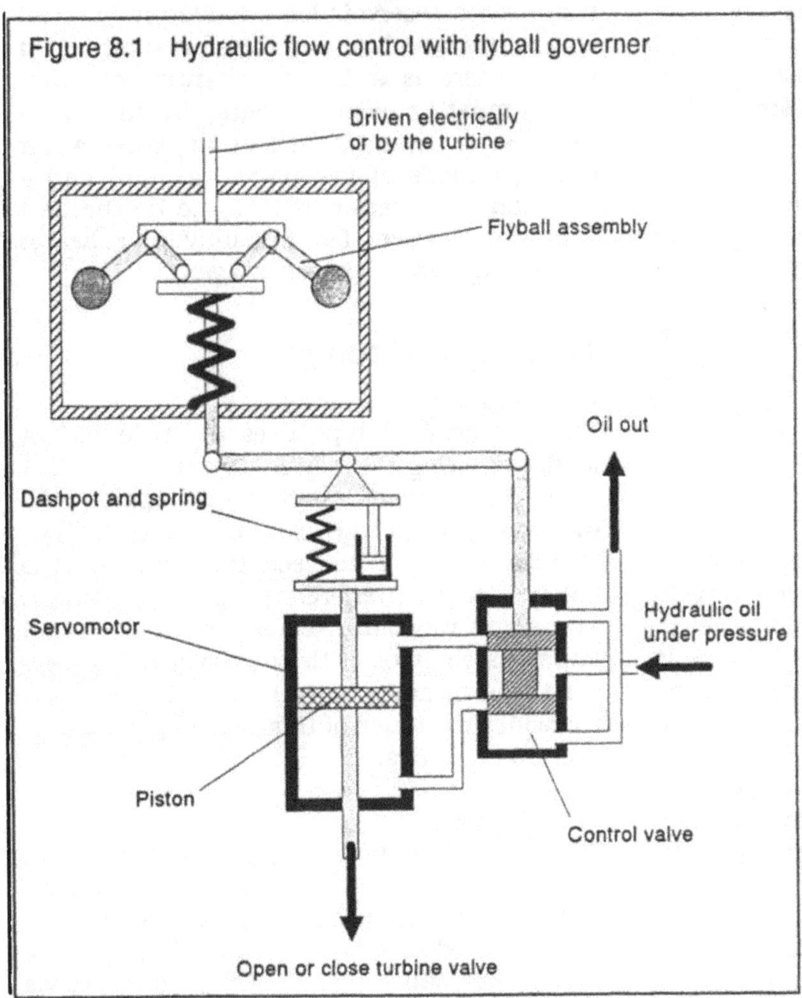

Mechanical flow control

Sometimes the force required to operate the flow control mechanism of a micro-hydro system can be small enough to make hydraulic control unnecessary. In this case the flyball assembly can be connected directly to the turbine valve, but the flyballs will need to be significantly larger to generate the required force.

As an alternative to a valve, a flow deflector slows down an impulse machine by partially (or wholly) diverting the flow away from its blades (see Figure 7.5). These governors are usually purely mechanical i.e. the output from the flyball arrangement is used directly to operate the jet deflector without the need for a servomotor. Some mechanical damping is usually incorporated to avoid instability. Such governors are fast and accurate and fairly simple to maintain, although in recent years they have been largely superseded by electronic load controllers.

There are other possible variations in the flow control system, for example electric motors to drive the flyball, or electric motors or actuators to move the turbine valves. These have the advantage of less complex maintenance but generally have a shorter design life and are less reliable.

Load control governors

As its name suggests, the load controller is a device which controls the electrical load on the generator, rather than the flow through the turbine. Its job is to maintain a fixed total load on the generator despite fluctuating user demand so that the turbine is held constantly at its design speed. It achieves this by employing ballast or 'dump' electrical loads to absorb the excess output of the alternator. If the demand falls to zero then the ballast load must be capable of absorbing the full output power of the turbine. The turbine can therefore run continuously at full flow and there is no longer a need for a flow-regulating mechanism and the complex system which that entails.

If electrical demand falls the turbine and generator speed up, so the output frequency increases. This frequency change is sensed by the load controller, and in response it switches in extra ballast load of sufficient resistance to dissipate exactly the same power as that which was switched off by the consumers, so keeping the total load on the generator constant.

The principal advantage of a load controller is that the overall system is less complex and less costly. It not only eliminates the need for a precision-engineered mechanical governor mechanism, but it may also allow the design of the turbine to be simplified because there is no need for the guide vanes or rotor blades to be adjustable since the turbine is always running at full flow. A less sophisticated system increases the reliability and reduces equipment costs considerably, especially for micro-hydro plants. For example, rather than a typical cost of US$10 000 for a hydraulic governor for an 8kW plant (perhaps equalling the cost of the turbine and generator), a single phase electronic load controller for the same plant could cost less than US$1000.

As they are solid state electronic devices, load controllers require no maintenance and have no moving parts. The disadvantage is that if electronic components fail in the field, they probably cannot be repaired there and then. However most units are composed of separate printed circuit boards which can be replaced easily. In fact established-brand electronic controllers have proven to be highly reliable.

A load controller also permits full exploitation of the energy generated. The energy absorbed by the ballast load is available in the form of heat and can be used to generate income e.g. by heating water or air, thus reducing the unit cost of the useful energy delivered.

When a load controller is used, the flow through the turbine is set at a constant level (but not necessarily maximum) and all the water is used for power generation. This is ideal for 'run-of-the-river' schemes where there is no reservoir for water storage. Load controllers are less suitable for schemes in which there is a limited quantity of water which must be used efficiently. In this case a reservoir is essential for storing excess water during periods of low power demand and a flow control governor is likely to be the best solution. However few micro-hydro schemes involve significant water storage.

Types of load controller

Proportional

The proportional type uses a single ballast load, adjusting the power diverted to ballast by various means, such as 'chopping' the waveform or, in the case of a DC ballast, rapidly switching between the load and the ballast. Radio interference can be a problem and some waveform distortion on the main supply is common with such systems. However these disadvantages have not restricted the widespread use of this type of governor in over 30 countries.

Step/block

The step/block type employs a number of separate ballast loads, usually with a binary distribution (e.g. 2kW, 4kW, 8kW) which can be connected in various combinations to give a range of ballast loads. Relays (sometimes solid state) are used to make the connections. Waveform distortion on the main supply is minimal with this system, but regulation is less accurate and voltage flickering, visible on lights, has been a problem. The ballast load wiring is complex and the values of the ballast loads are important, making replacement more difficult.

Three phase electronic load controllers
These units sometimes incorporate *phase balancing*. This means that the required ballast load is not shared evenly across the three phases, but is distributed in such a way as to compensate for unequal main supply loading. It therefore keeps the alternator phases evenly loaded at all times. Controllers without phase balancing require larger alternators and manufacturers will recommend the correct size. Although larger, they are cheaper and simpler and do not require bulky current transformers.

Manual control

Many small schemes throughout the world still rely on manual control to govern the power generating system, using either flow control or load control. However this is only really suitable for the smallest of systems having fairly steady electrical loads. It is not necessary for the operator to make flow or load adjustments continually during the operation of the plant, as long as the turbine speed remains substantially constant. The automatic voltage regulator (AVR) on the alternator will smooth out small fluctuations in voltage.

Flow modification
This approach requires a turbine that includes a flow-regulating valve. Load variation is observed on voltage or frequency meters to inform the operator that the flow to the turbine should be increased or decreased.

Load modification
This method can be used either to supplement flow control or with a turbine that has no flow-regulating valves. The manual modification of the load is similar in approach to that of an automatic load controller, but the ballast load has to be increased or decreased manually as the user load varies, as monitored by voltage or frequency meters.

Any manual control system is vulnerable to operator failure. A turbine is a very responsive machine, often capable of accelerating from an underspeed condition to full overspeed in a few seconds. Therefore for safety reasons, an ungoverned plant must have equipment capable of giving continuous operation at runaway speed (i.e. at full flow with no load).

8.4 Synchronizing with other Power Sources

Synchronizing with a diesel set

There are instances when it is useful to be able to run a diesel (or petrol) generator into the same load as the micro-hydro system, perhaps to increase the supply in the dry season. To do this the two alternators have to be synchronized, i.e. both are set to the same frequency, voltage and phase sequence. When they have been adjusted so that the output waveforms are identical in real time (i.e. so that they are in phase) they can be connected together. If they are not exactly synchronized, any significant instantaneous voltage difference between them will throw any safety switches ('trip' switches), blow a fuse, or cause damage.

Although the synchronizing process itself is quite straightforward, the subsequent behaviour of the system is rather complex. If the two machines are of similar output powers, then very careful design of the governing systems is required to prevent instability of their outputs. The generators used need to have compatible characteristics, otherwise large circulating currents may develop between them. The advice of their manufacturers should invariably be sought. If one machine is considerably smaller than the other then the control problems are reduced. However this will not always hold true; for instance a 10kW hydro can still overpower a 100kW diesel if there is no load on the system. In summary, it is dangerous to experiment without first consulting manufacturers or obtaining specialist advice. Diesel and petrol generator sets often have a very low tolerance of overspeed and are easily damaged if 'driven' by a hydro-turbine.

In control terms, for stable running as synchronized sets, the 'droop' (i.e. the pattern of voltage correction) built into the governors has to be very similar. This also applies to the AVRs (automatic voltage regulators) of the two alternators concerned. Most manufacturers supply different versions of their AVRs for synchronizing, and this will include a 'quadrature droop' circuit which prevents circulating currents in the system.

Assuming we have matched governors and AVRs, the equipment needed for synchronizing can be quite simple. Two single-phase machines with earthed neutral terminals can be synchronized with the help of two light bulbs. They are wired together in series and connected across the live terminals of the two machines. The reason for using two is that voltages equal to double the rated voltage will appear if the machines are exactly out of phase. The speed of one of the machines is then altered very slightly until both bulbs flash very slowly. Then, when both bulbs are completely out, indicating that there is no voltage difference between the two machines so they must be exactly in phase, a switch is closed joining the two alternators. A similar process using six bulbs is commonly used for three phase.

Load control governors usually allow fine tuning of frequency, which is then useful for manual synchronizing. There are various commercially available electronic 'black boxes', which automate the synchronizing process.

Induction generators can be synchronized into a larger synchronous alternator very easily, simply by connecting and 'motoring' the generator up to speed. In this way, a small ungoverned hydro output can be synchronized with a larger diesel system without AVR or governor problems. Reverse current relays can be used to prevent the diesel being driven by the hydro.

Synchronizing with the local grid

Micro-hydro plants are often connected to the local grid because this is a reliable load which provides a high load factor. In some countries, stand-alone sets are later connected into 'mini-grids' which will eventually be connected to the national grid.

Each electricity board has financial and technical regulations for interconnection to the loacal grid. Technical regulations are often very strict to protect maintenance crews from supplies 'backfeeding' into the system. It is essential to contact the local grid operator to discover precisely what the requirements are before planning to interface a micro-hydro system with the grid.

Drive Systems 9

9.1 Introduction

THE drive system transmits power from the turbine shaft to the generator shaft. It also has the function of changing the rotational speed from one shaft to the other when the turbine speed is different to the required synchronous speed of the alternator.

Figure 9.1 illustrates the use of a pair of pulleys with industrial V-belts, which is one of the simplest and most popular drive systems in micro-hydro plants. In this case the gearing ratio is set by the ratio of the diameter of the large pulley to that of the small pulley.

The easiest approach to obtaining the appropriate drive system is to buy a turbine and generator set from a single supplier, in which case the transmission will be designed and supplied by the turbine manufacturer as part of the system. In some cases however, it is necessary to match up a turbine with a separately supplied alternator. This requires a calculation of the power rating for the drive.

For example, an alternator rated at 50kW maximum output at a speed of 1500rpm is to be driven by a turbine running at 600rpm. Assuming the alternator is 70% efficient, what is the power to be transmitted by the drive and what should the required gear ratio be?

$$\text{Power transmitted} = \frac{\text{Alternator power}}{\text{Efficiency}}$$

Figure 9.1 A parallel shaft drive system using V-belts and pulleys

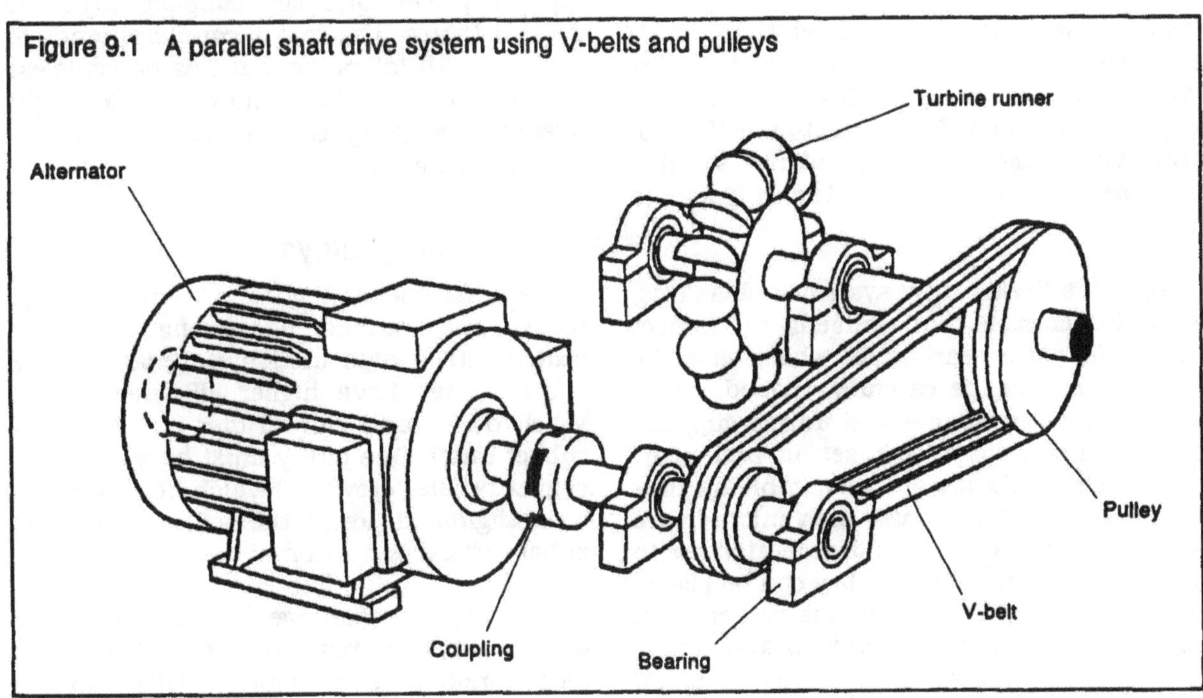

$$\text{Power transmitted} = \frac{50}{0.7}$$

$$= \underline{71\text{kW}}$$

$$\text{Gear ratio} = \frac{\text{Alternator speed}}{\text{Turbine speed}}$$

$$= \underline{2.5}$$

Once the power rating and gear ratio required of the drive have been calculated, a suitable system can be selected. This chapter summarizes the choices available. There is no general method for determining the size of the components that will be correct for a particular system because two of the important components, bearings and belts, have characteristics and properties which are unique to particular manufacturers (although most belt drives are offered in universal sizes and ratings). It is therefore essential to obtain manufacturers' catalogues for both bearings and belts (and any other special components such as seals or gearboxes), and to read the section which describes how to calculate the forces and determine the appropriate component sizes.

Rigid and flexible drives

It is usually convenient to have the turbine and generator shafts parallel and some distance apart, as in Figure 9.1. The simplest way to transmit shaft power is to use V-belts. This arrangement is particularly suitable because it can tolerate small misalignments of the shafts.

Instead of a flexible drive system such as this, it is also possible, but unusual, to connect the shafts via a gearbox, in which case the two shafts must be carefully aligned. Gearboxes are not recommended unless supplied as part of a hydro-electric set (in particular, old vehicle gearboxes are unsuitable as they lack the durability for use with micro-hydro systems). If the turbine and generator are to operate at the same speed, they can be placed so that their shafts are in line rather than parallel, in which case an industrial coupling is used to join the shafts and take up small misalignments.

9.2 Choosing a Drive System

The following options can be considered for micro-hydro drive systems:

- direct drive
- flat belt and pulleys
- 'V' or wedge belt and pulleys
- timing belt and sprocketed pulley
- chain and sprocket
- gearbox

Direct drive

This system is only for the case where the shaft speeds are identical because it uses a flexible coupling to join the two shafts together directly. The advantages are low maintenance, high efficiency (>98%), and low cost. The only disadvantage is that the alignment is far more critical than with an indirect drive.

Three examples of flexible direct couplings are shown in Figure 9.2. The diagram shows that misalignment can be both angular and positional. Couplings vary according to how much angular and positional misalignment they can tolerate. In general it is recommended that couplings with high tolerance be chosen, and preferably those which are composed of two separate pieces. One-piece couplings are more difficult to use. The cost of couplings tends to increase with tolerance, but it is nevertheless worthwhile spending more on a high-tolerance coupling then to suffer misalignment problems.

Flat belt and pulleys

Modern flat belts (Figure 9.3) run at high tension and are made of a strong inner band coated with a high friction material such as rubber. They have higher efficiencies than V-belt drives and run cleaner (i.e. with less rubber dust). One pulley must have a slightly convex profile ('crowned') which, together with good alignment, keeps the belt in position in either vertical or horizontal use.

The main disadvantage is that a higher tension is needed than with other drives (Two tons is not unusual) which means that the bearings suffer high loads, sometimes requir-

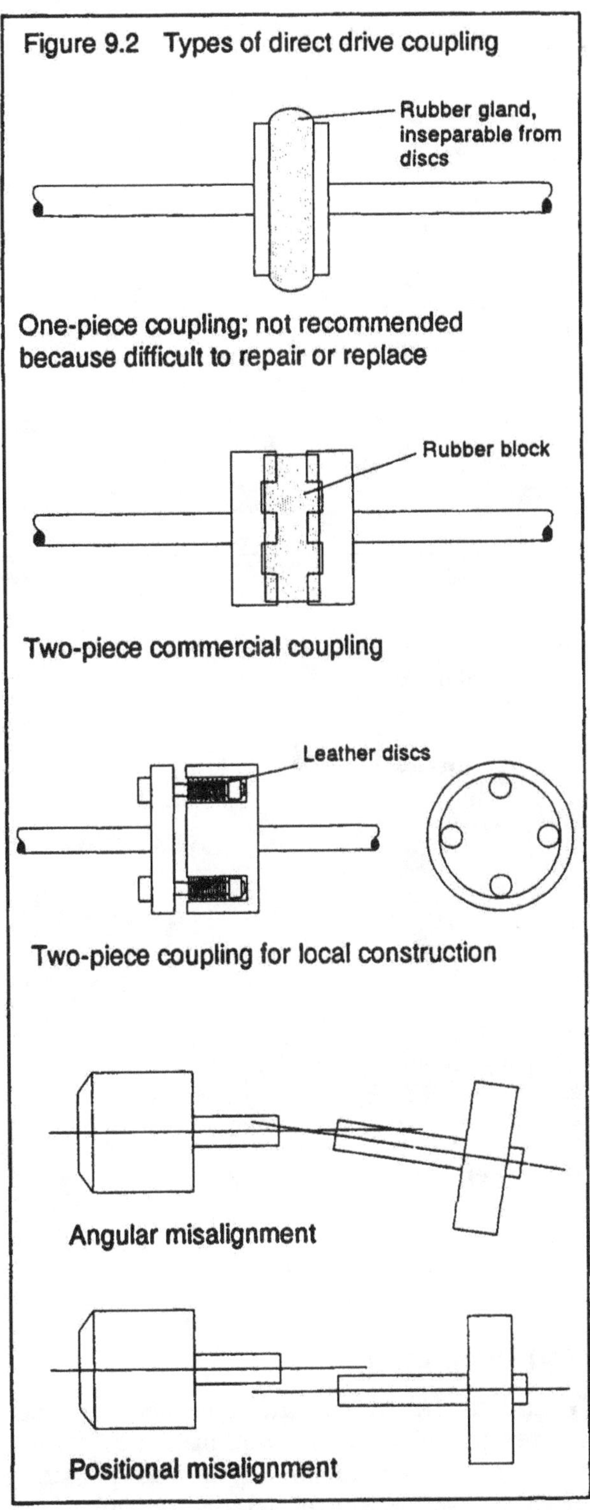

Figure 9.2 Types of direct drive coupling

ing additional layshafts to be used or standard alternators to be fitted with heavier duty bearings. Also their availability in some areas is less good than that of V-belt drives.

Flat belts generally require narrower pulleys than the equivalent multi V-belt, with advantages in cost and reduced overhang. Their maximum speed ratio is around 5:1.

'V' or wedge belts and pulleys

Illustrated in Figure 9.1, with a layout in Figure 9.4, this is the most common choice for micro-hydro sets up to 100kW. A major advantage is that these belts are very well known because of their extensive use in all kinds of small industrial machinery, hence they are also widely available.

V-belt manufacturers have developed industry standards for various types of V-belt drives, covering industrial, agricultural, and automotive applications. Only the industrial type of belts are of concern for micro-hydro use. Industrial V-belts are made in standard cross-sections and are referred to as *standard* (ordinary) and *wedged* V-belts. Ordinary V-belts are available in Y, Z, A, B, C and D cross sections, while wedged belts are available in SPZ, SPA, SPB, SPC and Delta.

V-belts differ from flat belts in that the frictional grip on the pulley is caused by the wedging action of the side walls of the belt within the pulley grooves. Therefore less longitudinal tension is required to maintain the grip and less radial load is imposed on the shaft and bearings.

Usually a number of V-belts are run side-by-side in multiple-grooved pulleys. Matched sets of belts are required to ensure even tension and these sets can be difficult to obtain in some countries. At higher powers and torques multiple V-belt installations can become cumbersome, with eight or more large belts and very wide pulleys.

The tolerance of misalignment of V-belts is very good, but efficiency is lower than other types of belt at around 85-95%. At very low powers the low efficiency can be a problem and timing belts are often preferred. Maximum speed ratio is around 5:1.

Timing belt and sprocketed pulley

These drives are commonly found on vehicle camshaft drives and involve toothed belts and pulleys. They are very efficient (about 98%) and clean running. The belt tension is lower than in any other belt drive, giving reduced bearing loads, but the belts do not slip on overload so cannot protect the shafts and bearings.

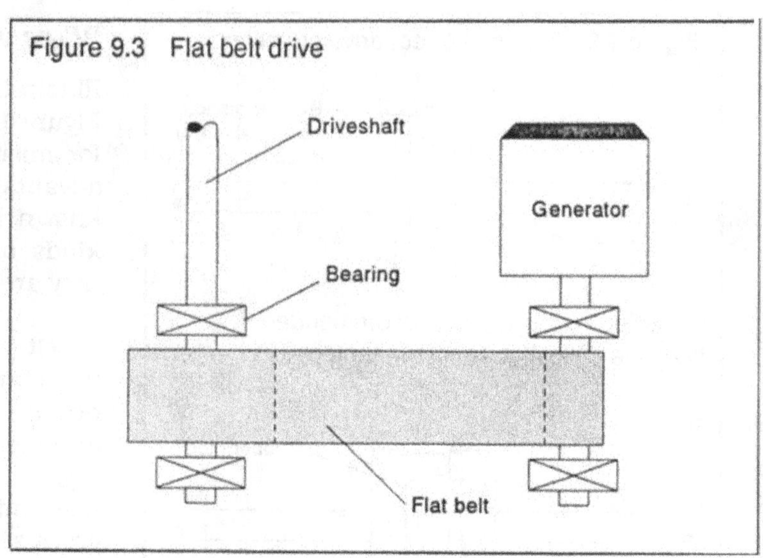

Figure 9.3 Flat belt drive

The main disadvantages are the cost of belts and pulleys and the low availability. They are especially worth considering for smaller drives (less than 3kW) where efficiency is at a premium. Speed ratio can be up to 10:1.

Chain and sprocket

Chains can have a very high efficiency but only at some sacrifice of lifetime. Long-life chain drives tend to be similar in efficiency to belt drives. Chain drives are not recommended because of their high cost, poor availability, the need to replace sprocket wheels periodically, and the difficult lubrication requirements. Very high speed ratios of greater than 20:1 can however be achieved.

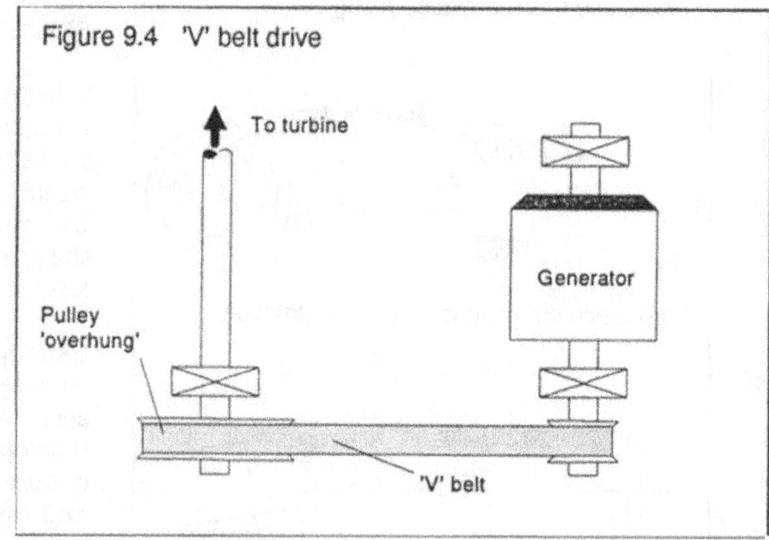

Figure 9.4 'V' belt drive

Gearbox

Gearboxes are used with larger machines when belt drives become too cumbersome and inefficient. Problems of specification, alignment, maintenance and cost rule them out except in cases where they are specified as part of a turbine-generator set.

9.3 Maintenance

Alignment

A common source of unreliability in drives is misalignment. If a system is inherently difficult to align then, although it may be set correctly on commissioning, problems are likely to occur during later maintenance. Particular points to note are to avoid shafts with three bearings (i.e. always use two) and to use self-aligning bearings wherever possible.

Direct couplings have specified allowable misalignments and these figures must be achievable on site, and regularly checked during routine maintenance.

Belt tensioning

'V' and flat belt drives have specified tensions. These are chosen to minimize slip whilst keeping bearing loads as low as possible. Modern flat belts are usually tensioned by drawing two parallel lines across the belt at a measured distance, and then tensioning until this distance increases by the specified amount, perhaps 4%, as measured with calipers and a steel rule.

'V' belts are often tensioned by 'feel' but this is an unreliable practice. Instead use hand-held belt tension indicators which allow rapid and accurate measurement of belt tension.

Electrical Power 10

10.1 Introduction

MACHINERY can be driven directly by a turbine, as in traditional corn mills and many modern timber sawing mills, but converting the power into electricity does have several advantages. For instance, it enables the use of all types of electrical appliance, from lighting to electric motors, and the flexible positioning of the appliances to wherever a power point can be set up, near or far from the turbine. A device which converts mechanical power into electricity is known as a generator. The most common type of generator produces alternating current and is known as an alternator.

Unless the supplier of the generator is providing the entire system, it is normally necessary to contract an experienced electrical engineer to design and supervise the installation of the electrical system, or at least to commission the system and ensure it conforms with national electrical standards. Nevertheless it is equally important that the engineer most closely involved with the scheme has a full appreciation of the following:

- the optimizing procedure to ensure a transmission line of minimum cost;
- the specification and purchase of electrical equipment;
- the selection and use of monitoring instruments such as frequency meters;
- the selection and use of switchgear;
- the proper practices which ensure safety including the selection and use of suitable safety devices;
- national electrical standards;
- commissioning, testing, monitoring, maintenance, and fault-finding methods.

These topics are covered in this chapter, with the exception of commissioning and maintenance which are covered in Chapters 12 and 13.

The design of a transmission line is taken first because it is often the most expensive single component of the electrical system, and the viability of the entire scheme may depend on the transmission costs. In addition, the electrical properties of the line (particularly the inductance) and the reduction in voltage along it will effect the choice and setting of the alternator.

The remainder of the chapter is largely concerned with the selection of a generator. Standard off-the-shelf generators are not designed for the particular conditions of micro-hydro, and are not perfectly suited to this application, for reasons which will be explained. Standard alternators do however have the advantage of being much less expensive than alternators custom built for micro-hydro installations.

Experience has shown that the best compromise in most cases is to buy a standard alternator whose rated power is 50% greater than the power expected from the micro-hydro installation. The alternator will then be capable of accommodating the particular duties imposed by a turbine but the cost will still be less than that of a specially built alternator. However it is always wise to consult the manufacturer of the turbine regarding the choice of generator. It is worth emphasizing again that it is preferable to purchase the turbine and generator together as a complete set covered by comprehensive supplier's warranties.

10.2 Electrical Theory

Basics

As a brief reminder of simple electrical theory, the flow of electricity or *current* (symbol I) is measured in amps (A). The *potential difference* (V) or 'pressure' of the electricity is measured in volts (V). The *power* (P), measured in watts (W), or more often kilowatts (1kW=1000W), is equal to volts x amps. The *resistance* (R) of a circuit is a measure of how well the electricity is being conducted (a poor conductor has a high resistance). Resistance is measured in Ohms (Ω) and is equal to the potential difference (or voltage drop) divided by the current. *Capacitance* (C) expresses the degree to which energy is being stored in an electric field rather than being available to do work, and *inductance* (L) is similar to capacitance but refers to magnetic fields.

AC and DC

Two types of current are produced by electrical generators, either alternating current (AC) or direct current (DC). In the case of AC, a voltage cycles sinusoidally with time, from positive peak value to negative. Because the voltage changes its sign, the resulting current also continually reverses direction in a cyclic pattern. DC current flows in a single direction as the result of a steady voltage. DC is not usually used in modern power installations except for very low-powered systems of a few hundred watts or less.

Alternating voltage can be produced in a stationary coil (or *armature*) by a rotating magnetic field (Figure 10.1b) but more usually a coil is rotated in a stationary magnetic field (Figure 10.1a). The magnetic field can be produced either by a permanent magnet or by another coil (i.e. an electro-magnet) known as a *field coil*, as in Figs 10.1c and 10.1d, which is fed by direct current (known as the *excitation current*). A generator supplying alternating current is described as an *alternator*, to distinguish it from a machine designed to supply DC current which is known as a DC generator or *dynamo*.

Current flows when a voltage difference is placed across a conducting body. In AC circuits the magnitude and timing of the

Figure 10.1 Alternator configurations

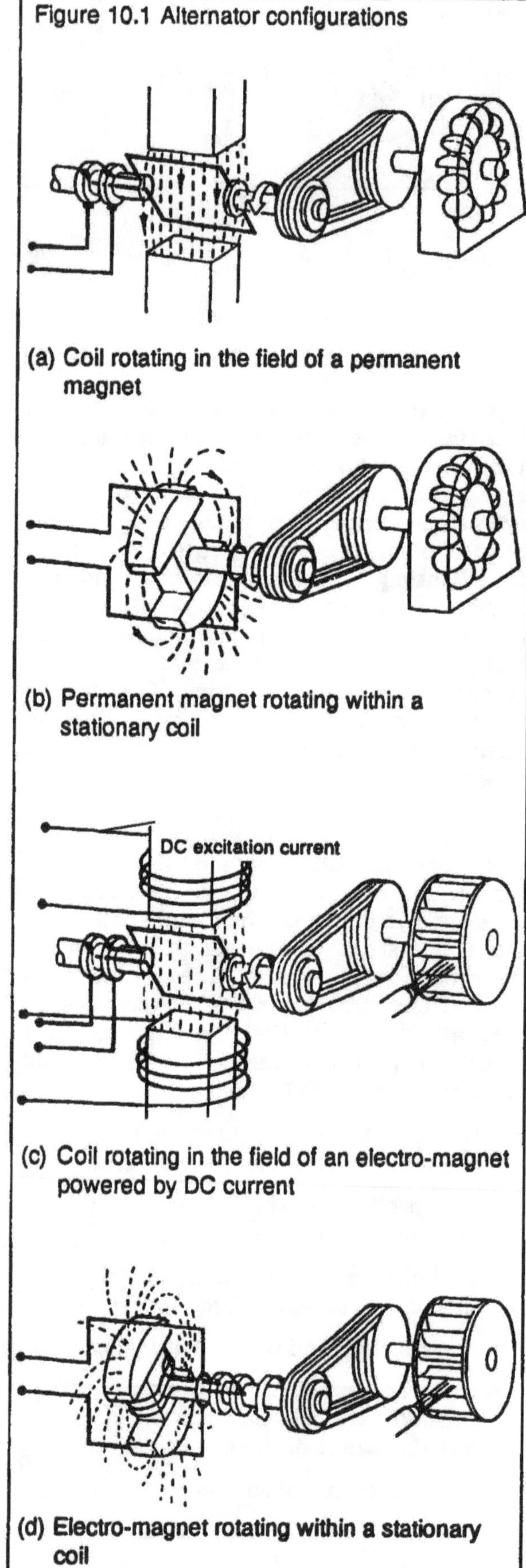

(a) Coil rotating in the field of a permanent magnet

(b) Permanent magnet rotating within a stationary coil

(c) Coil rotating in the field of an electro-magnet powered by DC current

(d) Electro-magnet rotating within a stationary coil

current cycle relative to the voltage cycle will depend on whether the conducting body is resistive, inductive, capacitive, or some combination of these elements.

The first three cases in Figure 10.2 show that for different types of load, the current cycle either (a) stays in phase with the voltage, (b) lags behind the voltage by a phase angle of 90°, or (c) runs ahead ('leads') by a phase angle of 90°. Circuits in which the load causes the current and voltage to be out of phase are said to have *reactive* loads. Generally a load is a combination of resistance, capacitance, and inductance, described by the term *impedance* (symbol Z), and causes a phase diffence between current and voltage of angle ϕ.

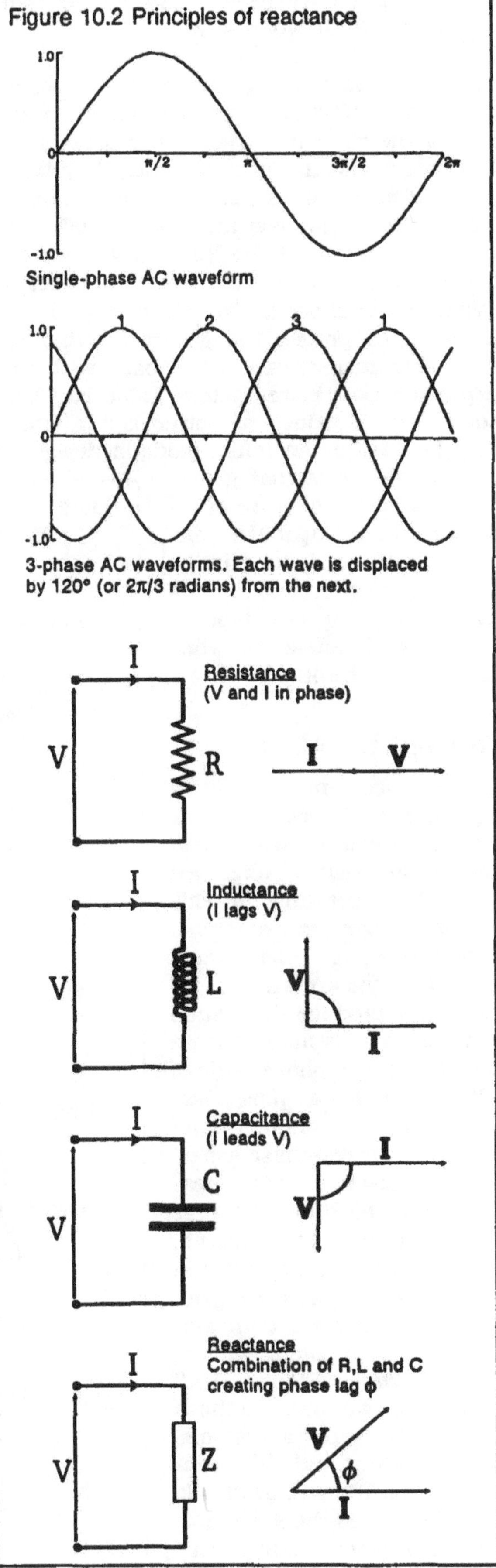

Figure 10.2 Principles of reactance

Power factor

Figure 10.2d shows the current-voltage characteristic of a circuit where the load causes the current I to lag the voltage V by an angle ϕ (the load in this case is predominantly inductive). Because the current and voltage are not perfectly in phase, the useful power available is reduced and is proportional to the cosine of the phase difference i.e. the power usefully consumed by the load is $VI\cos\phi$, although the power supplied is VI. The power not consumed is simply being shunted back and forth between supply and load. The ratio of useful power to total supplied power is called the *power factor* and is numerically equal to $\cos\phi$.

A typical electrical system might consist of a generator supplying a workshop with 10kW of useful power for lights and motors. The power factor of this kind of load could be about 0.8, implying a current lag of 37°. The supply power is therefore 10kW/0.8 = 12.5kVA. The convention is to refer to supply power as voltamps (VA) or 'apparent' power, while useful power, or 'real' power is in Watts. Apparent power is a misleading term because this power does actually exist and circulate in the system; it is generated and transmitted but it does no useful work. The problem is that the current carrying the power is larger than would be needed if the load was purely resistive with a power factor of 1. In the above example, if the voltage is 240V, the current will be VA/V = 12,500/240 = 52A. Had the 10kW load been purely resistive (e.g. all filament lights and no motors), the power factor would have been 1 and the current equal to 10 000/240 = 42A.

The total cost of the generator, transmission line and switchgear is predominantly a function of the current carried. Since poor power factors give rise to higher currents, they are to be avoided as far as possible. The transmission line itself can have an important effect on the power factor of the load. In general the generator is sized on the basis of a conservative estimate of overall power factor, and care is taken to protect the generator windings from over-current caused by loads with unexpectedly poor power factors. Table 10.1 lists some typical values of power factor which may be used as an initial guide in designing the system. Note that a load consisting of a mixture of inductive and capacitive elements will tend to have a better power factor than a load with only one type of reactance because the one counteracts the other.

Table 10.1 Power factors for electrical appliances

Appliance	Power Factor
Heaters	1
Cookers	1
Filament lamps	1
Motors, lightly loaded	0.4 lagging
Motors, heavily loaded	0.7 lagging

Three-phase AC

A single coil rotating in a magnetic field results in the single voltage waveform shown in Figure 10.2. An alternator operating in this way is a *single-phase* machine. Three rotating coils evenly spaced on the armature of an alternator produce three such waveforms displaced from each other by a phase angle of 120°; this is a *three-phase* alternator. It is common practice to use three-phase generation for power supplies greater than 10-20kW because three-phase generators, motors and transmission equipment are more compact and less expensive than single-phase equipment of the same power. Figure 10.3 shows the two main methods of wiring three-phase systems, known as *star* and *delta* configurations. The star system is often known as the *4-wire star connection* since it has a neu-

Figure 10.3 Three-phase AC systems: star or delta

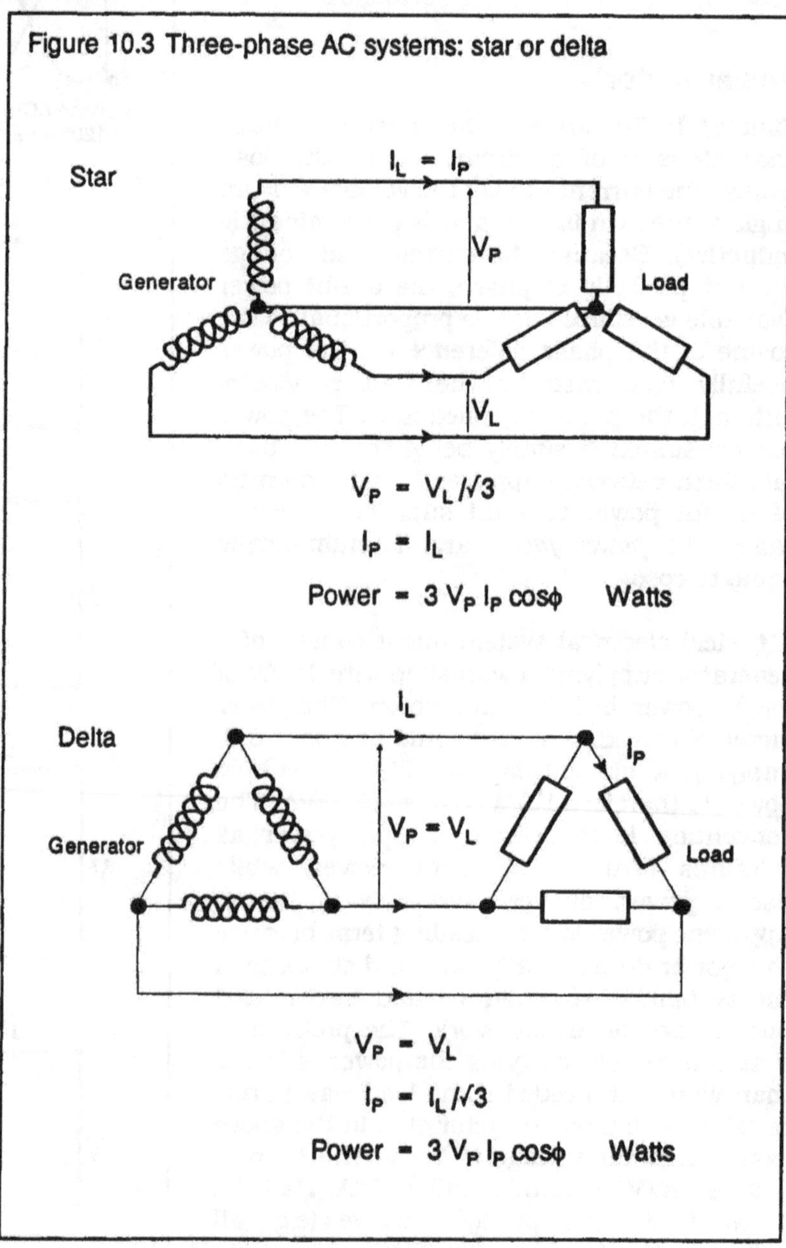

tral wire in addition to the wires carrying the three phases. The delta system is always a 3-wire system but does need to be well earthed at each end.

The load in a three-phase system can be connected either in the 4-wire star mode or in delta mode. The load in both cases is split into three parts, each connected to one of the alternator phases. In the star system the current drawn by each part of the load (I_P) is clearly equal to the current in the connected line (I_L). If each of the three loads has exactly the same impedance, then phasor diagram analysis (not covered here) shows that the voltage between two lines (V_L) is larger than the voltage across a phase (V_p) by a factor of $\sqrt{3}$. Exactly the reverse applies for the delta system. The kVA (apparent power) drawn by the combined load is the same in both star and delta systems. It is calculated as the product of the phase voltage and the phase current, multiplied by three for the three phases. The real power, in kW, is this power multiplied by the power factor of the load.

The advantage of three-phase systems is that power is delivered more efficiently and a three-phase transmission line is cheaper per unit of power delivered. The disadvantage is that the load must be split into three balanced parts. Imbalances in the load cause larger currents to flow. For this reason three-phase systems are rarely workable in electrical systems where the full generated power is below 10-15kW, since the switching in and out of individual loads will tend to unbalance the phases. This consideration rules out three phase as an option in many very small micro-hydro systems. A further factor is that switchgear and control equipment (for instance, over-current protection) tends to be more expensive in a three-phase system becuase it is more complex. Usually this last factor is offset by the following distinct economic advantages of a three-phase system:

- a three-phase transmission line requires less copper to transmit the same power, and so is less costly;
- three-phase motors are inherently smaller, cheaper and more easily available than single-phase motors of equivalent power;
- three-phase alternators are also cheaper than the single-phase equivalents.

10.3 Transmission Lines

Occasionally the turbine set is very close to the load, but more usually the load is remote from the most convenient powerhouse location and a transmission line is needed. The cost of electrical transmission depends mainly on the cost of conducting cable. This in turn is proportional to the square of the distance of transmission. For longer distances it is normal to increase the transmission voltage using transformers, as a higher voltage implies a lower current for the same power flow, which in turn permits more power to be transmitted for a given size of cable. There will be an economic maximum distance for a transmission line without transformers, typically around 1-5km for three-phase 440V lines. The main design criteria are:

- maximum allowable voltage variation from no load to full load;
- maximum economic power loss;
- protection from lightning;
- structural safety of overhead power lines e.g. in high winds or ice and snow;
- safety of operators, maintenance staff, and people living and working near the lines.

Design procedure

The main aim is to work out the size of conductor required, and hence the approximate total cost. If this is too high, either in absolute terms (cash available) or in comparison with the cost per kilowatt of the rest of the plant, then a higher voltage drop, or a higher voltage system (using transformers) may be considered. Although a higher voltage reduces the current and allows smaller conductors to be used, it also means that underground cables will require better insulation. Together with the transformer costs, the extra insulation will reduce the savings made on the cable thickness.

The main parameters which affect the size of the conductor are the choice between underground or suspended cables, the power factor of the load, the voltage drop along the transmission line, and the material used for the conductor. Such details as pylon spacing, insulators, and lightening protection can be decided afterwards.

Underground or overhead

Overhead lines are used most often because air is a simple and cheap cable insulation, but they are more vulnerable than underground lines, exposed to lightning, wind and falling trees. The land close to the lines has to be deforested (sometimes involving considerable compensation) and then cleared periodically. The pylons also have a finite life and need replacing approximately every 15 years. The wider the conductor spacing, the greater the inductive losses, so overhead lines are less efficient for a given conductor size than underground armoured cable, but bare uninsulated cable is more readily available in developing countries than underground cable, and is also cheaper.

Underground conductors have to be heavily insulated and are protected with steel braid, which tends to raise their cost to about three times that of overhead cable. Great care must be taken to anticipate problems such as ground movement, ploughing, and construction of new buildings. Once installed, however, the line should run without maintenance until the insulating material deteriorates, usually longer than 50 years.

As with penstocks, a buried transmission line is less visually intrusive and for that reason is probably preferable on environmental grounds, but any installation faults are less easily detected and corrected, so special care is needed in supervising the installation of underground cables to ensure the job is correctly and meticulously carried out.

Erection of overhead lines

The reader is referred to engineering handbooks (such as *Kempe's Engineers Year Book*) for design and practical guidance on overhead line erection. It is worth checking any proposed design against the following list of constraints:

- clearance between conductors, and from conductors to buildings, trees, etc.
- pylon spacing in relation to cable sag
- pylon life
- wind loading
- snow and ice loading
- conductor creep
- lightning
- corrosion
- maintenance costs
- line tensioning.

Determining the load power factor

The higher the power factor, the lower the current for a given power.

For single-phase:

$$\text{Current} = \frac{\text{Power}}{\text{Voltage} \times \text{Power factor}}$$

For three-phase:

$$\text{Current} = \frac{\text{Power}}{3 \times \text{Voltage} \times \text{Power factor}}$$

(where voltage is measured phase to neutral)

Resistance heaters, infra-red devices, storage heaters, filament light bulbs, and cookers, all have a power factor of 1. Motors have lagging power factors, typically 0.8, but as low as 0.4 if lightly loaded. Transformers also have lagging power factors, in the range 0.8 to 0.95.

Lagging power factors on factories are often 'corrected' i.e. improved from say 0.8 to 0.95 by the use of power factor correction capacitors. This is usually a good investment but often outside the control of the hydro engineer.

The power factors of existing loads can be measured with a kWh meter and an ammeter. Simply run the system with a steady load for, say, 10 minutes, and record the kWh units and the time in hours. Also measure current (I) and voltage (V). Then:

$$\text{Real power (kW)} = \frac{\text{kWh units}}{\text{Time (hours)}}$$

$$\text{Apparent power (kVA)} = VI \text{ (single-phase)}$$
$$\text{or} \quad 3VI \text{ (three-phase)}$$

$$\text{Power factor} = \frac{\text{Real power (kW)}}{\text{Apparent power (kVA)}}$$

This will give an accurate answer so long as the load is steady and balanced (if three-phase), and if the kWh meter is accurate. If there is any uncertainty, run the test with a purely resistive load and check that kVA = kW. Alternatively it is also possible to buy special power factor meters to do the job directly.

Allowable voltage drop

This is a critical design parameter because a slight change in value can greatly affect the cost. A typical figure for maximum permissible voltage drop along a transmission line given, for example, by the electrical utilities in the UK, is 6%. This is 14V on a 230V supply, and is a good starting point for the performance of a typical hydro transmission line. The best rule to follow is to run the alternator at a voltage 4% higher than the nominal voltage of bulbs and motors, and design the transmission line to give a voltage drop between 4% and 10%. The most economic balance between power loss and cable cost will determine the final voltage drop between these limits.

Conductor material

For overhead lines the choice is between aluminium and copper and is generally made on grounds of cost. In the exceptional cases of low power (less than 2kW) or long distance (more than 500m), pylon costs dominate, and stronger conductors will allow wider pylon spacing. Copper is stronger and cables incorporating steel strands, or even galvanized steel wire, can also be considered. For low-powered single-phase installations, co-axial armoured cable can be the least-cost solution for relatively small transmission distances.

10.4 Alternators

The choice of an alternator for a micro-hydro scheme can present difficulties. Off-the-shelf alternators used in diesel-driven generating sets are cheap and readily available, but precautions must be taken if they are employed in hydro schemes. Most diesel-driven alternators are designed to be mounted in line with the diesel engine, so that large radial loads are not imposed on the bearings of the generator. The bearings are therefore likely to be undersized for hydro applications in which pulley-and-belt drives are used. However the manufacturer may be able to supply the equipment with heavy-duty bearings on request. A second requirement of hydro equipment is the ability to withstand periods of overspeed. Some windings on the alternator rotor may be pulled out of position by the centrifugal force at an excessive speed, causing severe damage. The manufacturers of low-cost machines may not guarantee the security of the windings. If you are unable to obtain a machine guaranteed for overspeed, the options are:

- incorporate an overspeed trip which mechanically disconnects the generator;
- specify overspeed protection on the turbine e.g. jet deflectors tripped by a frequency sensor.

In general, modern alternators are designed to be most efficient when running at 66% of full rated power, so this is a further reason for selecting an over rated generator for micro-hydro applications. In summary, the chosen generator may sensibly be rated at about 50% higher power than the turbine's maximum output.

10.5 The Automatic Voltage Regulator (AVR)

The frequency and magnitude of the voltage produced in the output windings of a simple generator will vary with the shaft speed. Allowable voltage variation may be set at ±6% of rated value and frequency variation at ±3%, i.e. the frequency must not be allowed to rise beyond 51.5 Hz in a 50 Hz system. Where purely resistive loads are concerned, voltage variation can be +6% to -15% and frequency can be allowed to vary by ±25%. Outside these limits the performance and reliability of electrical appliances are seriously affected. The production of electricity within these limits demands some form of voltage regulation and for this purpose a control system is built into the generator, known as the *automatic voltage regulator* or AVR.

AVRs designed for diesel-driven generators which have lighter duty cycles have tended to be highly unreliable in micro-hydro installations, although the situation is improving. Continuous operation under conditions of varying load or shaft speed has led to frequent AVR failures, and the engineer is well advised to keep a stock of spare parts, especially rectifier sets and diodes. A diode expected to have a 2-year life in an intermittently operated diesel-driven generator may only have an 8-month life in a permanently running micro-hydro generator. The use of a thyristor-type electronic load controller may increase the electrical loading on the AVR and further shorten its life.

One useful method of prolonging AVR life is to reposition the AVR by removing it from the generator, extending the wiring, and bolting it to a neighbouring wall where it is protected from temperature variations, condensation, and mechanical vibrations. An ELC governed machine will run continuously at full load and will stay hot, so condensation at least will be avoided.

Another problem for AVRs in micro-hydro plants is that accidental low-power underspeed running of the turbine and generator may occur for prolonged periods. The AVR will respond by boosting the excitation (i.e. the strength of the magnetic field set up by the field winding) in order to raise the output voltage, but it cannot do this continuously without overheating. The high field currents forced in this way have been known to destroy the generator, even on under-speeds of only 5%. A number of methods of protection are therefore commonly used.

- Connection of the AVR to a shaft-mounted speed switch ensures that excitation is only applied when shaft speed is above 80% or 90% of normal running speed.
- Purchase of an 'intelligent' AVR which, for instance, reduces excitation with increasing output frequency. This is called *frequency roll off* and is illustrated in Figure 10.4.
- An alternative version of the above is to include a frequency trip in the circuit; a time delay on underspeed is needed to avoid undesired tripping on motor starts.

In general the intelligent AVR is the recommended method of protection. When purchasing a generator it is best to specify an AVR capable of full protection against under or over deviations in speed, voltage, current, and temperature.

Note that it is usually cheaper overall to buy a more expensive and sophisticated generator with an intelligent AVR than to buy a simple one and add trips.

Some AVRs will tend to give a voltage 'droop' on increasing load (i.e. the voltage level is not sustained at rated value but temporarily dips below it). This effect is worsened by increased transmission line losses, resulting in insufficient voltage levels at the load end. This characteristic is typical of the more old-fashioned 'wound' (or 'transformer-type') AVRs. These do not use semi-conductors and are very robust, so are often specified on micro-hydro sets despite their less accurate voltage control.

It is worth making careful contingency plans against generator breakdown. The failure of an AVR could result in long delays before the system is back on stream. Normally a plentiful supply of spare parts will be sufficient, together with immediate re-ordering of the stock when any of the spares are used. Since generators can be the 'Achilles heel' of a hydro system, it may be prudent in some situations to purchase two generators when setting up the scheme, allowing immediate replacement of a failed generator at any time. Repairs on the failed generator can then be made without withdrawal of the electricity supply.

Figure 10.4 Frequency roll-off

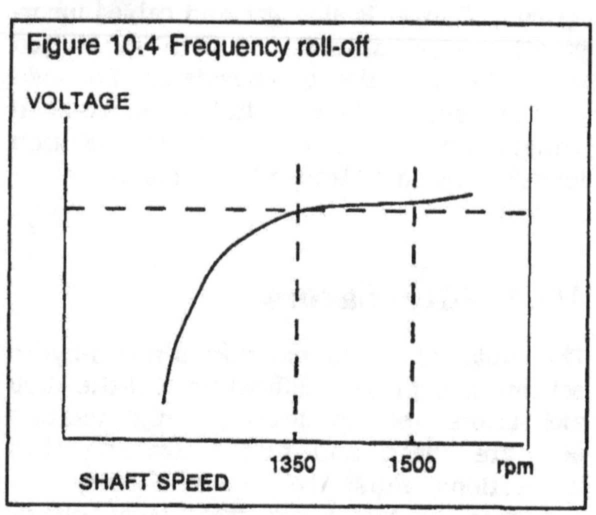

10.6 Generator Power

Section 10.2 introduced the distinction between real power and apparent power. The generator must be sized to produce the full apparent power drawn by the load and transmission, not just the continuous real power consumed. For this reason it is essential to specify to the generator manufacturer what this apparent power is expected to be. The best way to do this is to specify the maximum output power of your turbine in kW, since this is the maximum input power to the alternator. In addition an estimate of the power factor of the loads must be made so that the expected generator current can be calculated.

If you are in doubt but think you have a typical loading which is mainly inductive, it is common practice to specify a power factor of 0.8 lagging. However, on top of this, the power factor of the transmission must also be taken into account.

The example below illustrates why the power factor is so important to the generator manufacturer. If you underestimate it, the generator will not be able to handle the current drawn by the load, and will overheat. The greater the apparent power, the greater the current drawn, and the larger the generator winding needs to be.

Example to illustrate power factor and generator current.
Given a load rated at 10kW and 0.8 lagging power factor, what will the current supplied by an alternator be if it is a 240V single-phase machine?

$$\text{Apparent Power} = \frac{10\text{kW}}{0.8}$$

$$= \underline{12.5\text{kVA}}$$

$$\text{Current drawn} = \frac{12.5 \times 1000\text{VA}}{240\text{V}}$$

$$= \underline{52.1\text{A}}$$

Suppose we now include the effect of the transmission line, which includes a transformer, so is inductive, and will perhaps reduce the overall power factor to 0.7. What current must the alternator now supply?

$$\text{Current drawn} = \frac{10 \times 1000}{0.7 \times 240}$$

$$= \underline{59.5\text{A}}$$

For comparison, had the load been purely resistive (power factor 1), then the current drawn would have been only 42A.

10.7 Switchgear

The electricity generating system of a micro-hydro installation must include some form of switchgear to provide a measure of control over the electrical power flow and to isolate the power supply if necessary. It must therefore also employ protection equipment which can detect a variety of different faults and automatically activate the appropriate switches. Listed below are the names and functions of some of the common switches used on micro-hydro installations. These are available for both three-phase and single-phase systems.

1. **Isolators** are manually operated switches which isolate the load from the supply. They are relatively cheap but are rarely used because they offer no form of protection to the power supply.

2. **Switch fuses or main switches** are manually operated switches similar to the isolator but with the addition of a fuse on each phase conductor which protects against excessive current.

3. **Moulded case circuit breakers (MCCBs)** are now replacing switch fuses because they are cheaper. The over-current function in the MCCB is handled by bi-metalic strips which isolate the load by activating the circuit breaker. Possible additional features on the MCCB include control trip coils to trip the circuit breaker, magnetic trips to protect

against short circuits, earth fault relays to be connected to the trip coils, and motor drives to switch the circuit breaker on and off.

4. Oil or air circuit breakers are fairly old and are being replaced by MCCBs because they offer the same facilities at a much higher price.

5. Earth leakage circuit breakers (ELCBs) are a special case of the MCCB which can protect the circuit against earth leakage (see Earth Fault Protection below).

6. Contactors are generally used in control circuits such as in motor starters. However they can be activated by various trips to perform as very expensive circuit breakers.

7. Change-overs are used to select a source of power and channel it to a load. In micro-hydro systems a change-over may be used to connect the load to the main grid, or to the micro-hydro, or to another independent power supply.

10.8 Protection Equipment

Two types of protection are considered here: the protection of people from accidents and the protection of machines from faults. In a micro-hydro installation, electrical generation and distribution systems are protected using the trips listed below, in conjunction with either an MCCB or a contactor to isolate the circuit electrically. Note that all except the lightning trip will usually include a time delay so that they will not trip under transient conditions.

Trips

1. Under-voltage trips will trip the circuit breaker if the voltage drops below a pre-set value, normally an adjustable range down to −15%. Long periods of under-voltage can damage electric motors or the alternator itself and may occur in a micro-hydro installation for the following reasons:

→ defective AVR
→ machine over-loaded
→ load power factor is very poor
→ bad regulation on a transmission line
→ lack of adequate flow to the turbine
→ faulty ELC or turbine govenor

2. Over-voltage trips will trip the circuit breaker if the voltage goes over the pre-set value. Over-voltage can damage load appliances, and in extreme cases, may cause a breakdown of certain insulating materials in the circuit. It is generally caused by:

→ defective AVR
→ leading power factor load
→ bad regulation on a transmission line
→ faulty ELC or turbine governor

3. Over-current trips and other current-limiting devices are the most common form of protection used in electrical systems. Over-current damages the generator windings, switches, cables, and other equipment due to the excess heat generated in the conductors.

The simplest forms of over-current protection are fuses, followed by MCCBs. The third form of protection is an over-current trip. These different current limiters will isolate the load from the generator if it is drawing current in excess of the pre-set value. It is also possible to use PTC thermistors as a current trip. Over-current is caused by:

→ excess load on the generator
→ faulty equipment connected to the generator
→ lagging power factor
→ short circuit
→ incorrect alternator frequency
→ phase imbalance of a three-phase system

A fuse will have a range of protection up to a value of 1.6 times its rating, and an MCCB will have a value of 1.2 times its rating when protecting against over-current. It is important to protect each individual phase of a three-phase system.

4. Frequency trips will disconnect the load if the supply frequency is over or under the pre-set value. If supply frequency is not within its limits, it may damage the alternator and any electric motors drawing current. Under-frequency may be caused by:

→ overloading the machine
→ insufficient flow to the turbine

- → defective governor
- → belt slip

Over-frequency may be caused by:

- → excess flow through the turbine
- → defective governor
- → defective alternator or AVR
- → faulty ballast load

5. Temperature trips are activated when the winding temperature exceeds a safe value. The trips are wired to temperature sensing probes embedded in the alternator windings. The winding temperature may rise due to:

- → overloading the alternator
- → incorrect frequency;
- → excess current due to a low power factor
- → waveform distortion due to the load
- → defective bearings
- → inadequate alternator ventilation
- → high ambient temperature

6. Lightning protection is vital to prevent serious damage to the transmission line from either a direct or an indirect lightning strike. Direct hits are rare and the only protection is to install an earth wire over the transmission line. Lightning discharge nearby can induce high voltages in the transmission line which might damage electrical and electronic equipment connected to the line. To prevent this, lightning arrestors are installed on the line to divert these high voltage spikes to earth. The lightning arrestors have a very high resistance at normal voltage but if the voltage rises over a safe value the insulating material breaks down and conducts to earth. Once the high voltage is removed the arrestor returns to its original state. Metal-oxide varistors work in a similar manner but dissipate less energy.

7. Miscellaneous. Other trips are available, though not widely used on micro-hydro installations, as follows:

- phase failure trip, to protect against loss of one or two phases;
- phase sequence trip, to monitor whether the sequence of one alternator is correct when synchronized with a second;
- reverse power relay, to prevent power from one alternator flowing into a second synchronized alternator.

Earthing (grounding)

It is necessary to earth (or 'ground') all equipment as this will give the zero voltage reference for all other voltages. Generally all metalwork is earthed by wiring it to an *earth electrode*. This consists of a galvanized iron pipe or a copper plate buried in the ground. In the past it was believed that the copper plate provided the best earth, but any long conductor buried in the ground can be as good or better.

The resistance between earth and the electrode should be less than one ohm, measured with a high ohmage or meg-ohm scale ohmeter, usually referred to as a *megger*. To take this reading, one terminal of the megger is connected to the earthing electrode and the other to a known good earth, or to a temporary earth which is oversized to ensure its resistance is low. The meter then reads the total of the two earth resistances, from which the required resistance value can be inferred to a sufficient degree of accuracy.

Experience has shown that a good earth can be obtained by driving into the ground a 2m long by 50mm diameter galvanized iron pipe until almost buried. Alternatively a 4m length of heavy duty copper cable can be buried between 1 and 2m deep. If these do not indicate less than 1ohm earth resistance, it is necessary to bury a second similar earth and connect it in parallel.

Earthing systems

There are various earthing systems that are used in conjunction with earth fault protection. In one such system the neutral at the source is connected to the earth. Usually this means that the star point of a three-phase supply is wired to earth. Similarly, for micro-hydro alternators the neutral point of the generator is earthed.

Earth fault protection

People can be protected from live wires and live parts with the help of barriers. However, if metal parts accidentally become live due to faulty insulation they must be isolated from the source. The earth fault relay (EFR) and the earth leakage circuit breaker (ELCB) will disconnect the power supply if such a fault occurs.

Commercially available EFRs have a variable setting for leakage currents and are used to disconnect the main circuit breaker if a fault is detected. ELCBs have a fixed setting and can be used down the line to disconnect only the distribution circuits during smaller faults, thereby maintaining a supply to the unaffected loads.

10.9 The Switchboard

Wiring between alternator and switchboard

The general practice for wiring is to follow the local standards for the distribution of electricity within buildings and factories. Wire sizes are taken from the relevant tables and the wires are drawn through suitable steel or PVC conduit. If the alternator is belt-driven, a flexible conduit is used between the alternator and the rigid conduit, in order to allow movement of the alternator for belt removal.

In general the cross-sectional area of the neutral cable can be one-third the size of the main cables. However if an electronic load controller with continuous voltage control is used, the neutral cable should be the same size as the main cable because it carries a fairly high current between the alternator and the ballast load.

The output from the alternator is wired directly into the main circuit breaker which controls the power flow. The protection circuit trips are wired to control this circuit breaker and switch it off in case of a fault. Sometimes there are a few control wires from the alternator which also come into the switchboard, e.g. for voltage adjustment and temperature sensing.

The power flow is fed from the circuit breaker into a heavy strip of copper known as the *bus bar*. The bus bar carrying power from the turbine and alternator is referred to as the hydro bus bar, to distinguish it from any diesel or grid sources nearby. The electrical loads are also connected to the bus bar. If no other source of power is available then the loads are wired directly to the hydro bus bar through a set of switched fuses or MCCBs. If there is a second source of power, e.g. main grid or diesel generator, then the two supplies are brought into a change-over switch whose output is fed through a switch fuse or MCCB into the load. The change-over switch then permits the selection of either power source.

Metering equipment

It is often necessary to know what is happening to the turbine with regard to the load, and meters can be installed for this purpose. A pressure gauge will usually be supplied with the turbine, and sometimes a rev counter as well, although this last task can be done electronically by a frequency meter on the generator. The switchboard should at least have a voltmeter to indicate the output voltage from the alternator and an ammeter to display the current being drawn.

On larger plants supplying three-phase power, it is usual to have the following: a voltmeter with a selector switch to read the voltage between phases and the line voltage, an ammeter on each phase, a frequency meter, a power factor meter, a kilowatt meter, a kilowatt-hour meter, and a DC ammeter to read the field current value.

Other equipment

It is sometimes necessary to install a heater with temperature control within the switchboard to keep it dry.

10.10 Micro-hydro for Battery Charging

Batteries offer a means of transmitting small quantities of electrical energy over greater distances than would be economically possible with cables. They are essential with the smallest of hydro-installations running to only a few hundred watts, where a cable more than a few metres long would be prohibitively expensive in relation to the value of energy being transmitted.

Batteries can also be used in conjunction with larger micro-hydro to provide a small amount of electricity for use beyond the economic reach of the local distribution network. For example if a light and a radio are needed 2km away, it may be cheaper and more convenient to use a battery, recharged at the hydro station and transported by vehicle. The cost of transportation then comes into the equation, but often a vehicle will be travelling between the hydro-plant and the user anyway. In such situations it would be usual for each user to have two batteries, one in use at any one time and the other at the hydro-installation being recharged.

There are a number of manufacturers of very small hydro systems for charging batteries, most of them American. Some of the available systems are claimed to operate at heads of less than 1 metre, although 5-10m head is said to be preferable. Figure 10.5 is a schematic view of one available system which requires only a hose-pipe to supply the running water and costs about US$400.

When used with a conventional hydro-system, batteries can be charged using a standard mains voltage battery charger, as used by most garages. With small systems, it may be advantageous to recharge batteries at times when electricity is not needed for other purposes such as between midnight and 7am.

Batteries can also be used to provide short surges of higher power in situations where, say, only 100W is being generated, yet 200 or 300W is needed for short periods. With such small hydro systems electricity is usually generated as DC at either 12 or 24V for directly charging a battery.

Because batteries are direct current devices, and are usually of 12 or 24V ratings, they can only power low voltage end-use appliances, unless an inverter is used (an inverter is an expensive and sometimes inefficient device for converting DC to mains voltage AC). Fortunately many highly efficient low voltage devices do exist, including lights, radios and TVs, as well as small power tools.

Batteries

The best batteries for such duties are deep-discharge storage batteries. Car batteries are not really suitable because they only have a limited life if they are regularly deeply discharged. A specially designed deep-discharge battery, on the other hand, will tolerate regular cycling down to 50% or less of full capacity and should survive up to five years or more if properly looked after.

Batteries need to be maintained by topping up with distilled water occasionally and by ensuring they are never excessively discharged. The latter is best achieved by being reasonably generous in sizing a battery for a given duty (i.e. by not trying to get more out of a battery than is reasonable). Battery manufacturers and dealers can advise on the maximum recommended capacity of different batteries.

Figure 10.5 One available configuration of a battery-charging hydro system

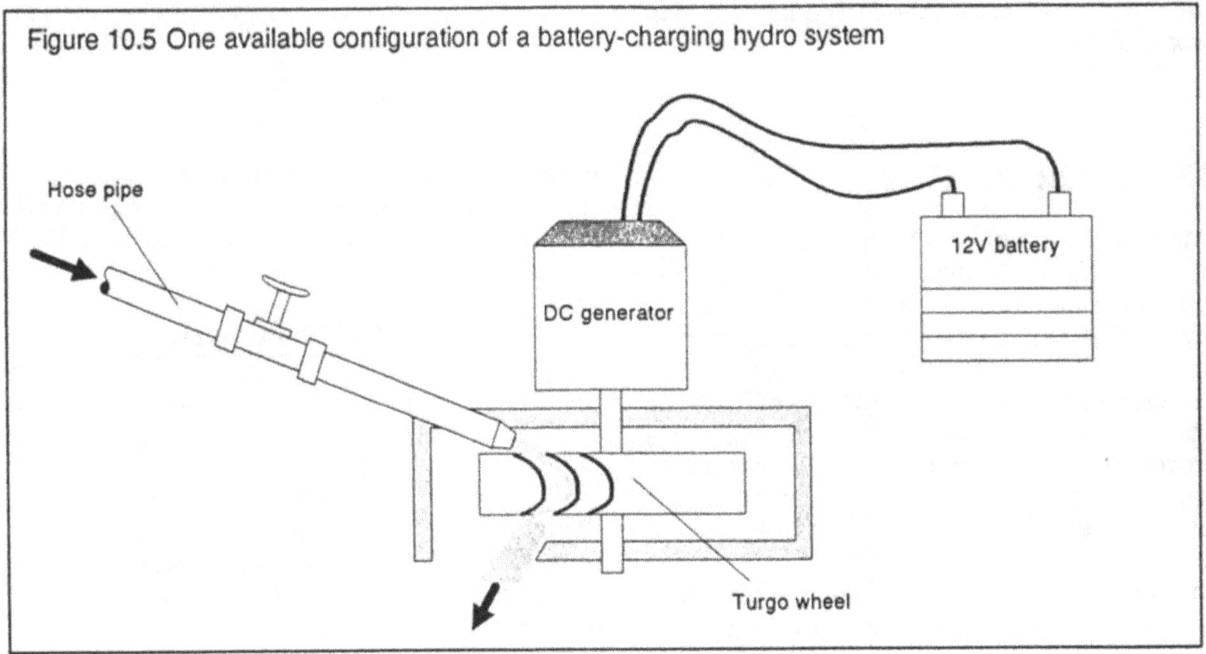

Basic Economics 11

11.1 Introduction

A hydro scheme necessarily involves considerable financial outlay. Table 11.1 gives a typical breakdown of the installation and commissioning costs. There will then be further expenditure on operation and maintenance, including the salaries of the operators and the cost of replacing worn components. In return the scheme provides useable energy, generally electrical power.

Having designed and costed the hydro scheme on paper, it is still important to consider whether other methods of providing the electrical power would be more sensible in economic terms. Diesel generation of electricity, for instance, is usually a far less costly option in terms of initial capital expenditure. The running expenses are higher, but these can usually be paid for by the revenue collected from the sale of the electrical power. In contrast the capital for the hydro scheme must usually be provided by a bank and paid back over a period of time in the form of loan repayments.

11.2 Discounting

Any sum of money can be invested so that its value increases in accordance with the current interest rate. If instead the sum is invested in the hardware of a hydro scheme, its potential for accumulating interest is lost. Consequently, the borrowing of money against future repayment is an expensive process, because the bank will expect to be compensated for both the original capital sum and the predicted accumulation of interest.

The prediction of the future value of a sum of money is known as *discounting*. In Table 11.1 the full capital cost of the fictional 40kW hydro scheme is US$90 000. This sum may be borrowed from a bank in return for annual repayments over a period of, say, 10 years. The bank may not necessarily require identical payments each year, but it is convenient in the financial appraisal of a hydro scheme to assume that the same sum will be paid each year throughout the loan repayment period. This is known as a *constant annual repayment sum* or the *annualized capital cost*.

In order to compensate for the loss of interest, which might be assumed to occur at an annual rate of 10%, the repayment has to be considerably greater than $9000 per year for the ten year period. If the discounted annual repayment is $A, the original capital sum is $C, the loan period is n years, and the interest rate is i (where i=0.1 for a 10% interest rate) then it can be shown that:

$$A = \frac{Ci(1 + i)^n}{(1 + i)^n - 1}$$

In this case the assumed interest rate is 10%, the repayment period is 10 years, and the capital sum is $90 000, giving:

$$A = \frac{90\,000 \times 0.1 \times (1.1)^{10}}{(1.1)^{10} - 1}$$

$$= \underline{\$14\,650}$$

Therefore the bank will expect an annual repayment of $14 650 over a period of 10 years, making a total of $146 500 for the original $90 000 loan.

Table 11.1 Capital cost example (40kW plant)

Activity	US$
Pre-feasibility study	500
Feasibility study, funding applications	1000
Design study	
Detailed surveys, production of tender documents, specifications, legal aspects, planning permission	2000
Site preparation	
Access tracks, clearing, temporary accommodation, etc.	2000
Civil works (installed costs)	
Intake weir	4000
Headrace channel	1000
Settling tank	1500
Channel	3000
Forebay tank	2000
Penstock + supports	20 000
Powerhouse, tailrace	2000
Electro-mechanical works (delivered to site)	
Turbine-alternator set	20 000
Load control + ballast tank	5000
Powerhouse, switchgear, protection	2000
Overhead line to demand point	4000
Installation costs	6000
Other costs	
Insurance, water abstraction license	1000
Supervision of contractors	3000
Contingency (10%)	8000
TOTAL	90 000

Note: This table is purely an example; in practice the above figures may vary widely depending on local circumstances.

11.3 Load Factor

It is now possible to calculate the cost of each unit of energy (or kWh) produced by the hydro plant over the 10-year period. To do this requires knowledge of the *load factor*. The load factor is the ratio of the actual energy used to the total energy that would be delivered if the system ran continuously at its rated power, i.e. it expresses the extent to which the hydro installation is actually exploited for profitable use. It can best be illustrated by way of a simple example.

Suppose that a 12kW hydro scheme is installed to provide a village of 60 houses with 200W of lighting per house. In the event only 30 houses are wired into the scheme, the remaining 30 householders having decided to wait a year and see how things go before paying their connection expenses. Lighting is only required on average for 6 hours each day throughout the year. Therefore in the first year:

$$\text{Load factor} = \frac{\text{Energy used}}{\text{Energy available}}$$

$$= \frac{6 \text{ hours} \times 200\text{W} \times 30}{24 \text{ hours} \times 12000\text{W}}$$

$$= \underline{0.125}$$

If in the second year the remaining 30 householders are connected, the load factor is doubled to 0.25.

The load factor is a strong indicator of the financial viability of the hydro scheme. A higher load factor indicates that the scheme can draw more revenue since only used energy gets paid for. If, in the above example, an industrial use can be found for the 12kW available power which uses electricity during 6 daylight hours, the load factor would rise to 0.5.

11.4 Unit Energy Cost

The unit energy cost of the electrical power supplied (in US$ per kWh) is given by the expression:

$$\text{Unit energy cost} = \frac{\text{Total annual cost}}{\text{Annual energy consumption}}$$

The total annual cost of the scheme is the discounted loan repayment figure (A) plus the annual running costs associated with operation and maintenance (O&M). The annual energy consumption is the installed power available in kW (P) multiplied by the number of hours in a year (8760), multiplied by the load factor (L):

$$\text{Unit energy cost} = \frac{(A + O\&M)}{P \times 8760 \times L}$$

The reader may like to confirm that if the 40kW scheme of Table 11.1 achieves a load factor of 0.6 while having O&M costs of around $10 000 per year, then it would have a unit energy cost of $0.12 per kWh.

The unit energy cost is the most useful indicator of the performance of the scheme. A very similar calculation can easily be made for alternative power sources, such as diesel generation and grid extension, and the different unit energy costs can be compared.

11.5 Payback Period

Another similar indicator of economic performance is the *payback period*, which is the time taken to 'break even', i.e. the period of time that elapses after the scheme is commissioned before the revenue collected is sufficient to pay off the original capital sum and the running costs to date. If the payback period is short, e.g. less than 4 years, the scheme is judged to be viable. Usually the payback period is calculated without respect to discounted future value of the capital invested, in which case it is referred to as the *non-discounted payback period*, or *simple payback period*. This is a useful general indicator, but it does require a reasonable estimate of the revenue expected from the sale of the electricity, which in some cases can be difficult to assess.

If annual revenue is expected to be R, and the original capital sum is C, then:

$$\text{Simple payback period} = \frac{C}{R - O\&M} \text{ (years)}$$

The annual revenue R is simply the energy sold multipled by the price it is sold at. If the selling price in $/kWh is S, then:

$$R = P \times 8760 \times L \times S$$

So, continuing the previous example, if the electricity can be sold at $0.15/kWh, then R = $31,536. Hence:

$$\text{Simple payback period} = \frac{90{,}000}{31{,}536 - 10{,}000}$$

$$= \underline{4.2 \text{ years}}$$

Thus the scheme would probably be considered to be economically viable.

Commissioning and Testing 12

THE final stage in the installation of a micro-hydro plant involves performance tests. The purpose is to check the function of the different components of the scheme and to measure the overall system performance against the figures arrived at during the design of the scheme.

12.1 Civil Works

The channel and penstock together represent a high proportion of the total capital cost of the hydro scheme. It is therefore of vital importance that they are installed correctly and are thoroughly tested before exposing them to non-stop river flow.

Channel

The channel should be thoroughly cleaned out to protect the turbine from the loose soil and stones that are likely to have accumulated. It may be easiest to let the water run along the channel with the flushing valve open for a good length of time, e.g. overnight.

The flow in the channel needs to be monitored to see whether it is behaving as expected. It is particularly important to find and repair any leaks along the channel and to observe the manner in which silt is settling along the route so that problem areas can be anticipated. The overflows need to be checked for size to ensure that surplus water will escape in a controlled manner and will never overflow the channel.

Penstock

If the penstock has been left empty for some time, there will invariably be stones and other debris inside. If possible it should be inspected when dry, but if not then water should be passed through it so that any objects will be carried to the bottom where they can be removed.

Debris can be removed fairly easily from impulse turbines because the penstock ends in an open jet, but where the penstock leads directly into the turbine casing, e.g. with a Francis turbine, removing unwanted objects can be difficult. These objects are also likely to do more damage to reaction turbines. Removing the inspection plates will give limited access to the turbine.

The penstock should be pressure tested by closing the main valve and letting the pipe fill with water. A 200% over-pressure test is usually specified for larger plants, following a procedure given by the manufacturer. It is important to fill the penstock slowly to allow the air to escape. There must be no air in the penstock during testing. Bleed valves may need to be fitted and opened to release trapped air. With small systems, holes can be drilled at high spots in the penstock and self-tapping screws fitted to close off the holes after the air has been released.

Particular care should be taken if second-hand equipment has been used in the penstock because it is much more likely to fail during the pressure test. Stay well clear until fully tested. It is also prudent to keep all power house doors and windows open when tests are carried out, because experience has shown that if a failure occurs the power house can be flooded within a very short time. It may also be wise, while constructing the power house, to ensure that all doors etc. open outwards so that they will tend to open easily if flooding occurs.

12.2 Electro-Mechanical Equipment

Before attempting to run the machines check that:

- all nuts and bolts are tight
- rotating parts are free to rotate
- bearings are lubricated
- couplings or drives are correctly aligned and tightened
- wiring for the alternator, switchboard, and ELC is according to the manufacturers' circuit diagrams
- ventilation is not obstructed
- the hydraulic governor is filled with oil
- the ballast load of the ELC is full of water
- all switches are in the off position
- the tailrace is clear

Alternator insulation levels should be checked with the use of high-ohmage meters (meggers) specially designed for this task. If the reading is low, the windings should be dried out in accordance with the manufacturer's requirements or as described in Chapter 13 on Operation and Maintenance.

It is always better to run up the system in stages, connecting the equipment progressively so that there is an opportunity to observe how each item is functioning. When doing so it is best to keep the main gate valve partially closed so that only the required amount of water is admitted to the turbine; it takes a long time to shut off this valve if anything goes wrong.

If possible, disconnect the turbine from the alternator by removing the drive belts, and run the turbine at a very low speed, e.g. 5% of the rated speed. If disconnection is difficult, run the turbine and alternator together with the exciter or AVR fuse disconnected; this will prevent the voltage from building up. The governor at this stage should be on manual control.

Look out for noise or misalignment in the drive. Any uneasy or knocking movement needs to be rectified. Check the bearing temperature against time to ensure that it stabilizes at a safe level. There may be oil pumps circulating oil into the bearings, in which case these should also be checked. After about half an hour of smooth running, raise the speed gradually in 10% intervals up to rated speed, making all the observations as before.

Mechanical governor

An advantage of mechanical governors is that they can operate on an unloaded turbine. Therefore after running at rated speed for some time without any problems, the governor, if it is a mechanical type, can be tested by putting it into automatic operation. The speed should stabilize at a fixed value determined by the setting of the governor. This may need some adjustment to bring the speed back to the rated value.

On flow control governors it should be possible to run the turbine up to 90% rated speed on manual control, so that when the governor is switched to auto-control it should increase the flow and raise the speed to the rated value. If the speed increases to a value higher than rated, the governor needs to be adjusted to bring the speed down. If the speed continues to increase, the governor is not functioning correctly and the setting-up procedure must be re-checked. Similarly, with the governor on manual the turbine speed can be raised to 10% above the rated value before switching the governor to auto-control; the governer should then reduce the speed to the rated value.

On flow deflection governors it may be more difficult to do such a test because of no auto-to-manual switch.

ELC governor

If the governor is an ELC (electronic load controller), it should be set up in accordance with the manufacturer's instructions, which will involve taking various readings against time once the alternator has been connected to the system. If the alternator is running into ballast it is necessary to keep the current in all three phases balanced and also take a reading of the neutral current. The machine should respond to adjustment of the frequency control of the ELC.

Alternator

If the mechanical governor is functioning correctly, and the insulation level of the alternator has been checked, then the alternator can be brought on line. Stop the turbine and re-fit the belt drive (or reconnect the exciter wires or AVR fuse). All connections must be carefully checked and double-checked against the wiring diagrams. The connection of monitoring and control equipment must also be inspected.

Perfect wiring is essential for a successful commissioning. If the alternator manufacturer has indicated a direction of rotation, the machine should run in this direction because only then will the alternator fan be supplying sufficient cooling air. It is important to ensure that the warm exhaust air from the machine is not drawn into the intake opening of the alternator fan.

Run the turbine up to about 5% of the rated speed and observe that the whole system is running smoothly. At this stage a small voltage may appear on the voltmeter. Gradually increase the speed to 100% on mechanical governing, but only to about 95% on ELC governing. This is because with ELC governing, as soon as the machine reaches rated speed the ballast load is switched on.

At the above alternator speed the voltage should remain steady, and with a three-phase system it should be the same for all phases. On an ELC system, if the machine is still running smoothly, now raise the speed so that it is just about rated speed. This should give a small ballast power.

AVR

On new alternators the AVR is factory set and it should not be necessary to alter these settings. However, if there is a need to change the voltage it is best to follow the manufacturer's instructions. If any changes are carried out they should be within the nameplate values. If they are not, the manufacturer should be informed of the final no-load voltage at the terminals, the no-load exciter voltage and current, and the full load values. This is important because if these values are beyond their design values it could damage the alternator or AVR.

Electronic AVRs will hold the voltage steady under the load condition to within ±2% or as stated in the specification. Transformer-type AVRs will have a variation nearer ±5%. It is also common for a transformer AVR to drop about 10V on phase-to-neutral when the machine is hot.

It is possible that an oscillation in the voltage will be noticed at this stage. It is important to locate the cause of this oscillation and rectify it. There are various possible sources; for example it may be due to the governor causing speed variation, it may be the belts slipping, or the AVR itself may be faulty, so it is important to check these out before attempting to change the settings on the AVR.

The alternator should be observed for some time before applying the ballast or main load. It is also important to record values of voltage, current, frequency, exciter current, temperature, etc. along with time of recording, ambient temperature and flow in the channel.

The water flow through the turbine can now be increased for loading the machine. If it is an ELC-type governor, the machine can first be tested with the ballast load. If it is a mechanical governor the machine must be tested on the main load or a dummy load. The loading procedure must be recorded and all readings entered against time.

12.3 Applying the Main Load

If a transmission line is included on the scheme this should be tested with a high-ohmage meter before energizing. All relevant wiring in the switchboard should be checked against the circuit diagrams.

On connecting the hydro power to the switchboard, the meters on the switchboard should indicate the presence of a stable voltage. The loads can be switched on individually and the currents recorded. If there are any electric motors, the direction of rotation of the motors should be checked. When heavy loads such as motors are connected it is advisable to wait a few seconds before starting a second load so that the turbine has time to stabilize after the first start.

Starting

It is sometimes difficult to start a small hydro plant if a lot of the appliances in the demand circuits are switched on, especially if much of the load consists of electric motors (such as in refrigerators or freezers) which draw a large starting current. In such cases the large starting surge can exceed the rating of the alternator and cause the system to trip every time the machine is switched on-line. In such situations the network needs to be broken down into smaller networks, with separate trips that can be brought on-line in sequence so that smaller starting surges result. Also, devices such as 'soft-starts' or 'time delays' can be fitted to appliances with large electric motors to reduce or delay the surge.

Final setting up of trips

Once the turbine and alternator performance has been verified, the trips can be set up so that the plant can be handed over to the client. The trips should be set up in conjunction with the test report because certain loads may take some time to start and, during starting, the turbine speed may drop below the trip preset, so indicating the need for a time delay. The trips should first be tested individually before testing them together. It may be prudent to initially err on the side of safety, i.e. letting the trips cut the circuit flow often; rather than under-setting them.

12.4 Summary of Readings

By taking the readings described in this section it is possible to work out the power available at site and all the power losses along the system. Individual efficiencies and the total efficiency can then be calculated. The readings may also be useful at a later date by acting as a set of reference data for the system. Any faults can be picked out from a new set of readings. For example if the alternator is running hot, the temperature can be measured and referred back to the original test report.

Flow measurements can be taken most easily in the channel or in the tailrace and used in the calculation of the gross hydro power available. If flow measurements are made in the channel, the channel flow can be adjusted so that all the water flows through the turbine without any overflow at the forebay.

Head loss in the penstock can be recorded from the pressure gauge at the base of the penstock for various flow rates. (Head loss = gross head - pressure reading in m).

Turbine speed (rpm) or output frequency (Hz) should be recorded for various loads and flow conditions.

Bearing temperatures should be recorded over a period of 8 hours against the ambient temperature.

Alternator speed should be recorded for belt-driven transmissions over a range of loads.

Alternator temperature at various points on the casing, and the inlet and outlet air temperatures, should be recorded with time against the ambient temperature.

Hydraulic oil temperature of hydraulic governors should be recorded over a period of 8 hours.

Alternator voltages and currents on the three phases and the exciter voltage and current should be recorded.

Power factor of the load and kWh meter readings should be recorded.

Alternator-to-ballast voltage and current for ELC governors is measured with special voltmeters and ammeters. They should read the 'true rms' values. The neutral current should also be recorded.

Transmission line power factor must be recorded, along with the voltage at both ends and the load currents.

Overspeed tests on the turbine and alternator are also required by some turbine manufacturers, in which case the performance data should be recorded.

Operation and Maintenance 13

THE importance of planned maintenance in the management of small-scale hydro schemes cannot be stressed too highly. The economic viability of a scheme depends on its reliability and any delay in obtaining spares can have a significant effect on income. This chapter considers the maintenance requirements of the various components of a hydro scheme.

13.1 Water Supply Line

Channel

The operator will need to inspect the channel and intake regularly. If the channel, settling tank, or forebay tank get partially obstructed by silt or debris (e.g. vegetation or stones) these must be removed. The trash racks may have to be cleaned daily to prevent leaves and vegetation from blocking the flow to the turbine.

With a well-designed channel system, this routine cleaning can be done without interrupting the flow of water. If the water does have to be stopped for cleaning, it should be done when the electrical demand is low. It is likely that this cleaning process will have to be carried out more often in the rainier periods of the year when more debris will be flowing in the river.

Settling tank

The settling tank must be cleaned out as and when it has collected too much silt. This can usually be done simply by opening the flushing valve at the base of the tank, but it may be necessary to stop the water to drain the tank and remove the silt with a shovel.

Forebay tank

In most hydro projects the penstock inlet in the forebay tank is covered with a fine-meshed trash rack to prevent leaves, etc. from reaching the turbine. These trash racks must be cleaned regularly to maintain the required flow to the turbine. Trash racks are best cleared with a purpose-made rake. Sometimes there is a series of trash racks which have to be removed and cleaned one by one.

Penstock

If correctly designed and built, penstocks need little or no maintenance, particularly when the plant is new. A monthly visual inspection is normally sufficient.

Joints need to be looked at for leaks and to see if they have been moved by thermal stresses. The supports need to be inspected both for damage and to establish that the drainage is adequate. There is always a danger that the surrounding ground will be washed away during heavy rain and cause the support to fail. Vegetation on the supports must be removed as soon as possible to prevent roots from cracking the cement work.

Steel penstocks need to be checked annually for signs of corrosion, and repainted every three years or so. Leaking penstocks should be repaired as soon as possible because the water escaping could erode the soil around the supports. All vent pipes must be kept clear.

13.2 Powerhouse

Maintenance and operation of the power house is a daily job which involves both mechanical and electrical work.

Gate valves

These normally require little or no maintenance other than the occasional application of grease or oil to the moving parts. Leaks on these valves can be repaired by putting new packing at the points where the moving parts are sealed. Sometimes a valve may not close due to an object stuck in the gate and the valve may have to be dismantled to remove it.

Turbine

The turbine supplier should supply a maintenance schedule. The tailrace needs to be examined regularly for obstructions and the turbine casing inspected for leaks. The bearings need to be checked daily. The bearing temperatures should be measured while running, to see if they settle at the normal temperature. Any new vibration or sound should be investigated as it may be due to a loose part. A screwdriver makes a good listening device for detecting bearing trouble. The sound should be smooth, and any grittiness or rumble indicates a deterioration.

Drive

Direct drives only require a periodic inspection to see if the nuts and bolts are tight. A flexible coupling needs more careful examination to verify that the flexible material is still in good condition. Bearings should be lubricated as specified in the manufacturer's handbook or, failing this, every 4000 hours (i.e. six months continuous operation), being careful not to over-grease or use contaminated grease. Belt drives need more regular maintenance to keep the belts at the correct tension and free from grease. Devices for measuring belt tension are available from belt suppliers and are recommended.

Generator

It is best to follow the maintenance instructions of the manufacturer, but if these are not available Table 13.1 may serve as a guide, indicating the frequency of maintenance activities for both a part-time and a full-time system.

Table 13.1 Maintenace schedule for alternator components

Maintenance work	8-hour operation	24-hour operation
Check noise and vibration of bearings while running. Check bearing temperature.	Monthly	Weekly
Check all terminals are firmly tightened.	Annually	Annually
Lubricate bearings.		
Grease:	Annually	6 months
Oil:	3 months	Monthly
Check bearings for radial play and noise.	Annually	Annually
Thoroughly clean, check windings and associated parts.	Annually	Annually
Wash out bearings and re-grease. Replace if necessary.	3 years	Annually
Brush gear (where fitted): check brushes, replace if necessary.	Monthly	Monthly
Slip-rings (where fitted): Check & machine if necessary.	Monthly	Monthly
Commutator (DC exciters only): Check & machine if necessary.	Monthly	Monthly

Insulation tests

It is important to perform insulation tests before first energizing an AC generator, or after it has been out of use for some time, because the winding bearing may be shorted or earthed. If machines have to stand for long periods without running, it is advisable to check the insulation resistance at frequent intervals. The insulation resistance of all windings should be tested with a 500V ohm-meter (megger). Beware: under **NO** circumstances should a test be carried out with a megger rated higher than 500V. Where high-voltage pressure tests are carried out, the diode assembly must be disconnected from both the field winding of the generator and the rotor of the brushless exciter.

On low-voltage machines up to 600V, a value of 1 megohm should ideally be recorded, and certainly not lower than 0.5 megohm. The machine must not be energized when the insulation resistance falls below 0.5 megohm.

It is important to be aware that the insulation resistance may fall dangerously low if the winding becomes damp during a period out of service.

If the insulation resistance is low, some source of heat such as strip heaters or electric lamps can be applied to the machine and the insulation resistance checked hourly until it reaches the desired value. Heaters and lamps may damage the windings if placed too close. The generator temperature must not exceed 80°C because excessive heat could destroy the diodes and winding insulation. Care should be taken to provide enough ventilation for moisture vapour to be able to rise away from the winding. If the windings are excessively damp, and the insulation resistance is very low, extra care must be taken in the initial stages.

Governors

Mechanical and hydraulic governors need regular maintenance. Their belts have to be tightened, oil levels maintained, and all mechanical parts inspected for freedom of movement. Any adjustment to the governor should be carried out by trained personnel with reference to the manufacturer's literature.

Electronic governors on the other hand require minimal maintenance. They should be kept dust free and there should be ample ventilation for the electronics. An occasional inspection of the inside should reveal the state of the electrical connections, and if there are any signs of corrosion they should be cleaned and tightened. Sometimes insects enter ELCs and cause problems. They can be kept out by screening the ventilation holes with mosquito netting.

The water supply to the ballast tank should be checked regularly for blockages and the tanks scrutinized for leaks. Scaling on the elements should be removed very carefully without damaging them. Defective elements should be replaced immediately. If trips are installed they should be tested to confirm that they function.

Switchboard and wiring

The wiring between the alternator and the switchboard should be inspected periodically for loose connections and for corrosion at the terminals which may cause overheating. All live parts should be covered and the earth checked for continuity. Switches should be tried out for good connection and if any signs of overheating are observed (such as discolouration or scorching), the switch should be replaced or repaired.

The working of trips should be tested periodically because they perform a crucial function in the protection of the equipment. Meters should also be thoroughly checked to ensure that the information they display about the operating condition of the plant is accurate.

Transmission lines

Overhead lines must be clear of trees as these will cause earth faults and short circuits. Insulators on the line need to be inspected periodically. Corrosion can occur where there is direct contact between copper and aluminium conductors and this should be avoided by using proper jointing devices.

Underground armoured cables need not be maintained but it is sensible to isolate and megger test the line annually to check for deterioration or damage.

13.3 The Operating Schedule

The operating schedule should consist of a list of tasks to be carried out periodically or whenever deemed necessary. The accomplishment of tasks should be reliably and clearly recorded in a log book. These tasks might include:

- switching loads
- recording meter readings
- visual inspection of components e.g. intake, channel, etc.
- routine operations such as cleaning the trashrack, flushing out silt basins, and greasing bearings.

The operating schedule must also include full instructions for starting and shutting down the system.

13.4 Planned Maintenance

Civil works

Maintenance of the civil works can be carried out on a register basis in which a schedule is drawn up showing what tasks are to be performed and how often, tabulated separately for normal conditions, dry season, and wet season. It is important to remember that most problems will occur during times of heavy rainfall when the system may have to cope with many times the average flow rate, so more frequent inspection will be necessary during the rainy season.

Repair work on the other hand is more easily carried out during the dry season. The maintenance of the civil works should therefore be planned to take account of these seasonal characteristics.

Electro-mechanical equipment

Maintenance of the electro-mechanical components is not so dependent on seasonal variations, but those jobs which require the plant to be shut down for a period should if possible be planned for the dry season when the available power is at a minimum.

Inventory of spares and tools

This is a vital and often neglected part of budgeting. For example, in a US$30 000 installation, a full set of tools and spares will cost around US$300, i.e. only 1% of the capital cost. Yet the price for not having the right bearing or the right tool in stock could be one month's downtime for remote sites. This might cost US$1000 in lost energy and a lot more in terms of the disruption of load activities. Therefore with any new system at least one complete set of spares and essential tools should be specified by the supplier and passed on to the client.

Each system must have a carefully thought-out inventory and a re-ordering system which minimizes the chances of downtime. An example is shown in Table 13.2.

Table 13.2 Inventory of spares and tools

Spares etc.	Qty	Re-order time	Supplier
Multi-purpose grease	2 litres	1mth	ADM
V belts,SPC,270mm	6 off	2mths	BRH
C2163 roller bearings	2 off	6mths	STC
CX2116 ball bearings	2 off	6mths	STC
6BX001 diodes	6 off	6mths	SS
AVR No 233	1 off	12mths	SS
High-speed fuses	3 off	6mths	BC
3m penstock length	1 off	9mths	OTT
Penstock paint	5 litres	2mths	FF
Multigrade oil	2 litres	1 wk	FF

Tools

1. Grease gun
2. 30mm ring spanner
3. 32mm open ended spanner
4. 8 piece metric ring spanner set
5. 6 piece screwdriver set
6. 38 piece socket set
7. Fenner belt tension checker
8. Screen rake
9. Spade
10. Oil can
11. Electric blower
12. Wire cutters
13. Pliers

13.5 Alternator Fault Diagnosis

The most common alternator faults, their possible causes, and the appropriate remedy are tabulated in Table 13.3 for ease of reference.

Table 13.3 Alternator trouble-shooting

FAULT: No voltage from the alternator

Cause	Remedy
AVR fuse blown.	Check and replace with identical type.
Residual magnetism is too small due to heavy shocks during transport or a prolonged period of standing idle.	With the alternator at rest apply a 12 or 24V DC supply to the field winding for a short time (the *battery test*). This should restore the residual magnetism. It is also possible to run the alternator with the battery still connected to the field.
A break or bad connection on the exciter wires, field windings, or main stator windings.	Check cable connections, exciter windings and field windings. Battery will not excite field winding.
Speed is too low.	Measure speed and bring up to rated speed.
Heavy load already connected to alternator before starting.	Check turbo-generator is isolated from load.
Exciter diodes defective and will not excite with the battery test above.	Replace diodes if necessary.
AVR defective.	Replace if necessary. If the alternator excites on the battery test, it is a clear indication that the AVR is defective.
Internal short in windings.	Battery test will not generate a voltage. Test windings.
Brush gear worn, or not free to touch rings.	Clean and replace if necessary.
Manual regulator defective.	Regulator open circuit.
Defective slip rings.	Megger test slip rings with all electronic components disconnected.
Field wires have to be interchanged.	If the field wires are connected to the wrong poles of the exciter or AVR,the alternator will not excite.
Short circuit on field surge protector.	The field surge protector is across the field winding to protect the exciter diodes against a high voltage.If this is short-circuited, the field current will not build up.

FAULT: Alternator voltage too low with no load

Cause	Remedy
Speed not correct.	Measure and correct.
AVR defective.	Check with battery.
Voltage pre-set too low.	Adjust preset.
External pre-set	Preset not connected or defective.
Single-Transformer AVR	Adjust air gap or tappings.
Three-Transformer AVR	If three transformers are used in the AVR it may be that one transformer is defective.
Manual regulator.	Stuck or open at some points.
Dirty brush gear.	Clean brush gear.
Brush gear moved.	Rotate brush holder to best position on commutator to get highest DC voltage.
Rotating diodes defective.	Disconnect and check.
Poor connections.	Improve connections: bad ones will be hot.
AVR sensing wires connected to wrong terminals.	If electronic AVR it may be sensing a high voltage, e.g. phase to phase instead of phase to neutral. If transformer type AVR it may be sensing a low-voltage, e.g. phase to neutral instead of phase to phase. Check circuit for low-voltage winding tappings.

FAULT: Alternator voltage too high with no load

Cause	Remedy
Speed too high.	Correct speed.
Voltage pre-set too high.	Adjust preset.
External pre-set	Defective or not in circuit.
AVR defective.	Replace AVR.
AVR sensing wires open.	Check wiring.
AVR sensing wires connected to wrong terminals.	If electronic AVR it may be sensing a low-voltage, e.g. phase to neutral instead of phase to phase. If transformer AVR it may be sensing a high-voltage, e.g. phase to phase instead of phase to neutral.

FAULT: Alternator voltage drops with load

Cause	Remedy
Speed drops with load.	Check speed with load.
Belt slip	Tighten belts.
Alternator overloaded.	Check current and reduce if necessary.
Bad power factor.	Try some other load.
If current sensing is available on the AVR it may be connected the wrong way.	In transformer type AVR's interchange the voltage sensing wires.
AVR setting is not correct.	AVR setting may be under-compensated.
Quadrature droop kit.	If wrongly connected, disconnect if unnecessary.
In the case of an alternator with a separate exciter, the load-dependent excitation is incorrectly connected. Thus the excitation decreases instead of increasing with load.	Compare the separate exciter unit with the wiring diagram. With a meter unit, it should be observed that this voltage should increase with load.
Brief voltage collapse when large loads are switched on.	A brief voltage collapse of 15% is allowed. If more than this it should be started in a different load manner, e.g. Star/delta.
A diode may be defective.	Check and replace.
Internal winding may be short-circuited.	This may not show on a simple battery test.
Brush gear worn-out.	Sparking may be seen on the brushes. If the brushes are pressed down the sparking should reduce and voltage should improve.
Unbalanced loads.	Check loads.
Severe waveform distortion.	Check type of load and change if possible.

FAULT: Alternator voltage hunts continuously

Cause	Remedy
Speed not stable.	Check governor and speed.
Belt slip.	Check belt and tighten if necessary.
Flat belts.	The flap of long flat belts could cause this. Adjust stability of governor or AVR.
AVR	Electronic AVRs have an adjustment for stability. In electro-mechanical AVRs, the springs need to be adjusted.
Bad connection.	A bad connection to the AVR can cause hunting due to sparking in the connection.
Slip rings worn-out.	Due to irregular surface on the rings.
Large loads.	Due to large pulsating loads.
External interference.	All alternator control wires should be screened.
Defective bearings.	Due to uneven air gap, check and change bearings.

FAULT: Neutral conductor too hot

Cause	Remedy
Various star points are connected to one another and large circulating currents of triple mains frequency are flowing in the neutral wire.	Connect star point choke to suppress circulating currents in neutral wire. Alternatively leave star points without connecting together.
A highly unbalanced load is over loading the conductor.	Measure neutral conductor currents and distribute load symmetrically on all three phases.
Type of governing.	If an ELC is used which has a waveform control type switching, then a neutral current will be present between alternator and ballast. Use a large cable.
Type of load, e.g. three-phase star connected tranformer.	It may be due to triple mains frequency currents in the primary of the transformer. Disconnect neutral and check.

FAULT: Alternator becomes too warm

Cause	Remedy
Inlet or outlet of the alternator cooling system is obstructed or partially throttled.	The inlet and outlet of the cooling air should not be obstructed.
The warm outlet air is free to enter the inlet openings for cold air, resulting in a re-circulation of warm air.	Clean mesh if necessary.
The ambient temperature is too high, caused by cramped space, lack of ventilation.	This may occur below the alternator, e.g. in the bed of a base plate. A re-circulation of cooling air must be prevented by means of partition sheets (baffles).
Alternator winding is badly contaminated causing cooling to be ineffective.	The power house must be ventilated; the ambient temperature should not be above rated alternator temperature.

Alternator overloaded or power factor too low.	Clean alternator with dry compressed air when at rest and clean windings. Reduce output current to rated value, correct power factor with capacitors.
Waveform distorted.	Due to a bad load, or the use of thyristors.

FAULT: Alternator runs roughly, bearings noisy or hot

Cause	Remedy
Alternator under strain and incorrectly aligned.	Re-align alternator.
Alternator has been transported with locked rotor causing 'brinelling' of the bearings.	Replace bearings.
Bearing has too little grease.	Replace grease filling immediately.
Bearing has too much grease and is overheating.	Alternators with automatic grease feeders do not suffer from excess grease, so this is only possible in machines not having this device. If the bearing temperature does not drop after a period of running, some grease should be removed from the bearings. The permissible bearing temperature increase is rather high, i.e. about 50°C, so at an ambient temperature of 25°C bearing temperature of 75°C is permissible.
Bearing is defective due to normal wear.	Install new bearing of same grade.
Main rotor has a short circuit between windings.	Test: alternator runs faultless without excitation on no load, vibrations appear and intensify with load. Measure the resistance of main rotor windings.
If the shaft or housing has been filled or sleeved.	The shaft diameter is too large and bearing fitted tightly on shaft. If housing diameter too small and bearing fitted tightly into housing.
Bearing locks or sleeves touching housing.	Check housing.
Bearing loose in housing.	Check housing and bearing for marks. If the bearing is free in the end-plate housing it will overheat.
Plain bearings – if bearings have been remetaled.	Scrape bearing to suit shaft.
Excess belt slip. The pulley overheats and the heat flows along the shaft to the bearings.	Tighten belt or replace.

FAULT: Alternator vibrates

Cause	Remedy
Loose bolts.	Tighten fixing and foundation bolts.
If alternator has been rewound the rotor is out of balance.	Remove belts and rotate alternator shaft, if the shaft returns to the same position it is out of balance.
The pulley may be out of balance.	Remove pulley from alternator and perform above test. Balance alternator with pulley.

BIBLIOGRAPHY

WMO Operational Hydrology Report no.13 (WMO no.519)

WMO Guide to Hydrological Practices Vol.1 (WMO no.168).

Mini-hydropower Stations: a manual for decision makers, UNIDO, New York 1983

A.Arter, U.Meier, "Harnessing Power on a Small Scale", *Hydraulics Engineering Manual*, SKAT, St Gallen 1990

A.Brown, "Stream Flow Measurement by Salt Dilution Gauging", ITIS, Rugby 1983

A.Derrick, C.Francis, V.Bokalders, *Solar Photovoltaic Products: a guide for development workers*, (Second edition), IT Publications, London 1991

G.Fischer, A.Arter, U.Meier, J-M.Chapallaz, *Governer Product Information*, SKAT, St Gallen 1990

J.Fritz, *Small and Mini Hydropower Systems: resource assessment and project feasibility*, McGraw-Hill, New York 1984

Hangzhou Regional Centre (Asia-Pacific) for Small Hydro Power, *Small Hydro Power in China: a survey*, IT Publications, London 1985

R.W.Herschy, *Streamflow Measurements*, Elsevier Pubs 1985

D.Hislop, *Upgrading Micro Hydro in Sri Lanka*, IT Publications, London 1983

R.Holland *Micro Hydroelectric Power*, IT Publications, London 1983

R.Holland, *Micro Hydro Power for Rural Development: lessons drawn from the experience of ITDG*, UNIDO 1983

A.Inversin, *Directory of Sources of Small Hydroelectric Turbines and Packages*, NRECA International Foundation, Washington 1981

A.Inversin, *Micro-Hydropower Sourcebook*, NRECA International Foundation, Washington 1986

D.Jantzen, K.Koirala, *Micro-Hydropower in Nepal: development, effects and future prospects with special reference to heat generator*, FAKT, Stuttgart 1989

H.Lauterjung, G.Schmidt, *Planning of Intake Structures*, GATE, Wiesbaden 1989

U.Meier, *Local Experience with Micro-Hydro Technology*, SKAT, St Gallen 1981

U.Meier, A.Arter, *MHP information package: A selected annotated and classified bibliography on micro hydropower development*, SKAT, St Gallen 1990

U.Meier, M.Eisenring, A.Arter, *The Segner Turbine: a low-cost solution for harnessing water power on a very small scale*, SKAT, St Gallen 1983

R.Metzler, H.Scheuer, R.Yoder, *Small Water Turbines for Nepal: the Butwal experience in machine development and field installation*, FAKT, Stuttgart 1986

D.McGuigan, *Harnessing Water Power for Home Energy*, Garden Way Publishing Co, Vermont 1978

M.Nilsson, S.Nilsson, *Mini/Micro Power Development in Zambia: report and description of the project*, SIDA, Sollentuna 1984

E.M.Wilson *Engineering Hydrology*, (3rd edition), Macmillan, London 1983

NEWSLETTER

'HYDRONET', an international newsletter for the dissemination of micro-hydro expertise, published by FAKT and SKAT.

USEFUL ADDRESSES

Deutsches Zentrum für Entwicklungstechnologien (GATE)
PO Box 5180
D-6236 Eschborn 1
GERMANY

National Rural Electric Cooperative Association (NRECA)
1800 Massachusetts Avenue
NW Washington
DC 20036
USA

Förder gesellschaft für angepasste Techniken in der Dritten Welt (FAKT)
Gänsheidstrasse 64
D 7000 Stuttgart 1
GERMANY

Swiss Center for Appropriate Technology (SKAT)
Varnbüelstrasse 14
CH-9000 St Gallen
SWITZERLAND

IT Publications
103-105 Southampton Row
London WC1B 4HH
UK

SKAT Bookshop
Tigerbergstr. 2
CH-9000 St Gallen
SWITZERLAND

www.ingramcontent.com/pod-product-compliance
Lightning Source LLC
Jackson TN
JSHW050909150125
R13932800001B/R139328PG77033JSX00001B/1
9781853390296